In Defense of the Alien

Bertram Russell's Fatalistic Prognosis for the
Far Future of Man and an Alternative Future

ALBERT DE ALIEN

In Defense of the Alien
Copyright © 2022 by Albert de Alien

All rights reserved. No part of this publication may be reproduced, distributed, or transmitted in any form or by any means, including photocopying, recording, or other electronic or mechanical methods, without the prior written permission of the author, except in the case of brief quotations embodied in critical reviews and certain other non-commercial uses permitted by copyright law.

Tellwell Talent
www.tellwell.ca

ISBN
978-0-2288-8168-1 (Paperback)
978-0-2288-8167-4 (eBook)

Table of Contents

Preface 1

Part 1 The Universe, Aliens and Human Evolution 3

Other life in the universe 3
The dangers of the universe and its alien population 5
The limitations of terrestrial-based evolution
of the human race 9
Self-directed evolution of aliens 10
Attempting to communicate with aliens 15
How humans might appear to aliens.. 20
How alien involvement in human affairs
might appear to humans 21
The goal of future human evolution 22

Part 2: The Human Soul and Brain Copying.................. 24

The human soul.. 24
Mapping or copying the human soul,
and its relevance to religion.. 25

Part 3: The Necessary Ethical Evolution of Aliens,
and Convergence on a Universal Moral Philosophy........ 32

The necessary ethical evolution of aliens................................. 32
The convergence towards a universal moral philosophy.......... 33
A long-term teleology for humankind,
and its present relevance... 42

Part 4: The Ethical State of Humans 48

The role of the human subconscious 48
Maslow's Hierarchy of Needs and
the drivers of human evolution 51
Self-interest and greed as evolutionary motivations................. 53
Altruism... 60
Sports and entertainment as the opium of the masses............ 62
The human sex drive...65
Physical beauty as a motivator... 69
Toxic masculinity and war... 71
Toxic masculinity and political and
economic manipulation 89

Part 5: Tribalism as Both a Motivator and
Consequence of Evolutionary Behaviour 92

Tribes and how they work... 92
Tribes built around conspiracy.. 96
Religious tribes 100
Wealth-based tribes...107
Nationalist tribes 114

Part 6: Economic and Social Inertia
of the Dominant Social Group......................................117

Part 7: Maslow's Higher-Level Needs:
Self-Esteem and Self-Actualisation 121

Part 8: The Role of Language, The Need for a Common Human Language, and The Need for a Language for Communicating with Aliens 124

The role of language 124
The need for a common language throughout humanity...... 127
Communications with the aliens 129

Part 9: Priorities for Humankind's
Immediate Future 132

Protecting the Earth 132
Expanding into the solar system 137
The path for the short-term future...140

Preface

One of humanities most famous mathematician, logician, and philosopher, Bertrand Russell (1872–1970) once wrote the following pessimistic and fatalistic prognosis and lament for the far future of humanity:

> That man is the product of causes which had no prevision of the end they were achieving; that his origins, his growth, his hopes and fears, his loves and his beliefs, are but outcomes of accidental collocations of atoms; that no fire, no heroism, no intensity of thought or feeling, can preserve a life beyond the grave; that all the labours of the ages, all the devotion, all the inspiration, all the noonday brightness of human genius, are destined to extinction in the vast death of the solar system; and the whole temple of Man's achievement must inevitably be buried beyond beneath the debris of a universe in ruins - all these things, if not quite beyond dispute, are yet so nearly certain, that no philosophy which rejects them can hope to stand. Only within the scaffolding of these truths, only on the firm foundation of unyielding despair, can the soul's habitation be safety built.
> ["Free Man's Worship" Russell B, 1903]]

Albert de Alien

Russell's prognosis is only conditionally true. There are alternatives for humanity, depending on choices that humanity makes for itself. On the other hand, Russell's prognosis is excessively optimistic in assuming that humanity will at least survive until the death of the Earth and the solar system. This is a possibility, but it is far from a certainty, and it says nothing about the state that humanity will be in at that time.

Part 1 The Universe, Aliens and Human Evolution

Other life in the universe

The lights that twinkle in the night sky are not just for the musing of human poets or to guide sailors. They also tell of a vast and timeless universe, interspersed with intelligent life— but some humans argue that life on the Earth is the only life in the universe. They principally point to the absence of any evidence for the existence of life elsewhere in the universe. However, these people should concede that their argument is flawed by reason of a simple logical principle: the absence of evidence of a thing, including life beyond the Earth, is not evidence of its absence. In any event, my presence is their contradiction.

Others argue that the formation of life is itself an inherently implausible event in the universe, and that the Earth is simply a highly improbably exception.[1] This topic is the subject of the nascent study of astrobiology and the search for simple life in the solar system and elsewhere. This argument is just an expression of human hubris and will hopefully soon be only of historical curiosity.

Historically, Judeo-Christian and Islamic religions have reinforced this hubris via the doctrines that God created humankind in God's own image, redeemed humanity from its fall, and has long-term plans for humanity—and for humanity alone—to join God in 'heaven.' This simple picture is doomed if it transpires that God created more than one intelligent species, with the same redemptive goal for each.

If the above hubris is true, then it follows that God created an essentially eternal and infinite universe just to engineer a single spark of life—one spark, on one planet, in one solitary galaxy at a single temporal point in the life of that eternal universe. If one subscribes to this premise, then God is an incredibly inefficient and wasteful cosmic engineer.

There are others for whom the stars represent both an invitation and a challenge, in order to develop a greater understanding of the universe which humankind can accept or reject at the peril of its own future. Fortunately, there are humans who do accept this challenge with passion and a career-long devotion. It is these scientists of various kinds upon whom the future of humanity depends, and to whom humanity will turn in desperate times.

The dangers of the universe and its alien population

Science fiction movies usually, if not always, emphasise the dangers of the universe for humans. This is obviously true, but only as a matter of degree. As any number of disaster movies portray, humanity also faces real dangers to its Earthly existence. What is portrayed in the cinema or on television is just a perspective of the universe as held by those who make the movies and their financiers.

Dominant human perspectives of the universe are governed by the human condition on Earth, the Earth itself, the solar system, and that which may be gleaned from telescopes. Humans understandably interpret the universe and its relative dangers from their familiar terrestrial perspective. This is a very limited perspective upon which to form any views about the dangers of the universe. Living in the Arctic or any of Earth's other extreme environments is dangerous. It is less so for the Inuit, Yupik, and other Indigenous peoples who inhabit these environments.

Science fiction movies and television series typically present aliens as a danger that exists in the universe. Aliens are often portrayed as large, violent, ugly, and technically superior invaders. The iconic television series *Dr Who* and *Star Trek* are good examples, as are the movies *Alien* and *Independence Day*. The purpose and intent of aliens is characteristically portrayed as the subjugation or extermination of humankind and/or the plundering of the Earth's resources. Humans are frequently portrayed as hapless, passive, righteous victims who resort to violence only as a means to resist the evil invaders.

Fears about aggressive aliens are not new. They were famously featured in Orson Welles' radio broadcast of H. G. Wells' *World of the Worlds* in 1938, which led to widespread panic and a belief that aliens were taking over the Earth. Scientists have also attempted to quantify alien risk through the 'Rio Scale,' and the United Nations has adopted a non-binding protocol to deal with any first contact with aliens.

Eminent physicist Stephen Hawking was so concerned about the risk of an alien invasion that he publicly challenged the right and the logic of the METI program to direct radio signals into space in its quest to make first contact. He likened METI signals to the risk of a person 'shouting in the jungle' and attracting predators. He pointed to cases in human history where militarily stronger societies have encountered weaker ones, leading inevitably to near genocide. Hawking saw humankind facing the same threat, from aliens responding to METI signals. He queried METI's apparent assumption of a global mandate to take on this risk, not only on behalf of humanity, but for the Earth itself. Hawking was not alone in this concern. Another famous physicist, Freeman Dyson (1923–2020), expressed the same concerns.[2]

METI has been functioning for around 50 years. Its signals have therefore extended to a sphere with a radius of 50 light years. In this space, there are about 2000 star systems with their own planets. Of course, METI signals are not the only means by which aliens could detect humankind on Earth. Humans have been leaking light, radio, and televisions signals into space for over a century, and there are about 15,000 star systems with their planets, within this volume. If there is merit in the criticism of METI, then consistency would demand that humans cease all domestic communications that cause signal leakage from the Earth.

METI, SETI, and electromagnetic leakage are not the only means by which advanced aliens could detect the presence of human civilisation. Aliens could detect and visit the Earth by any number of means, without humans even being aware of this.

The process of anthropomorphising

The process of attributing human motivations and behaviour to other, non-human life forms is known as 'anthropomorphising.' It also typically involves attributing the worst aspects of human nature to the other life forms. Human motivations and behaviours are a product of human needs. Human needs are a product of their physical, psychological, and social evolution on Earth. It is not logical to project those needs and motivations onto species that are not human, have not undergone the same evolution as humans, and therefore cannot have human (or human-like) needs.

In human terms, aliens have every right to feel insulted by these human projections, in much the same way that humans would reject and feel offended by comparison of their nature to that of worms, or other barely sentient life forms.

The first fundamental proposition: human thinking about aliens is mostly anthropomorphic. Human are projecting the worst of human nature onto their representation and understanding of aliens.

To avoid the danger of conscious or unconscious anthropomorphising, humanity needs to realistically assess its own evolution and its present state. A realistic assessment is not one that is a mere apology for the human race. If realistically conducted, such an assessment is likely to cause offence and incite defensive reactions in many humans.

But because humans are not likely to undertake this task for themselves, I have commenced the exercise for them in this small book.

The limitations of terrestrial-based evolution of the human race

The Sun is a middle-aged star. The Sun and the Earth are about 4.5 billion years old. Life on Earth has had several false starts that were largely exterminated through various partial-extinction events.[3] But for these extinction events, the evolution of life on the Earth would have almost certainly been different to the present form of life on Earth. The human form is not a necessary outcome of evolution.

Planetary catastrophes causing total or partial extinctions are events that may or may not happen. They are not necessary events in the history of a planet. Hence, not all evolving life in the universe has experienced a resetting of its evolutionary clock due to extinction events. Some life forms have had the total life of their planet in which to evolve.

The present human state is partly due to genetic factors and partly due to socialisation, both of which are part of the wider evolutionary processes. Early human evolution was Darwinian in nature. This is a blunt evolutionary instrument, rooted in competition for survival, and 'red in tooth and claw.'[4]

But the process of human evolution has itself evolved over time, induced by a growth in learning. Humanity's increasing knowledge base, in turn, causes an increase in its capacity to acquire further knowledge. The learning process for all species, including humans, is an accelerating one, analogous to Moore's Law.[5] It is for this reason that the intellectual paradigm shift during the Scientific Revolution, from the 15th to the 17th centuries, was so pivotal to human development. The resulting increase in knowledge fostered further adaptions and/or modifications to the environment, and further minor adaptions of the human form itself, which will accelerate with time.

Self-directed evolution of aliens

The acceleration of the evolutionary process for all surviving species is caused in part by the onset of what is known as self-directed evolution. This is an accelerating process following a form of Moore's Law. Self-directed evolution is an important concept and a process which occurs progressively over deep time—the time scale of geologic and cosmic events. It occurs when a species has sufficiently adapted to its physical environment and developed some capacity to choose its preferred environment and its preferred physical form. This development includes all necessary and consequential sociological and ethical growth.

For those species in the universe that have not suffered extinction events (or at least not as recently as those on Earth), they have had the opportunity for the self-directed evolution of their technology, sociology, and their preferred physical form over billions—not just tens of thousands, or millions—of years, far more than humans. This is not to say that continuous progress in any evolutionary process is guaranteed, and that restarts do not occur.

Humanity may be a latecomer to self-directed evolution, but it has begun. There are examples of this in genetic engineering, in various medical sciences, in planetary research, and in information technology and artificial intelligence. Assuming humanity survives its present and near future, then the potential exists for human self-directed evolution to continue into the far future.

The second fundamental proposition: human evolution is undergoing an accelerating process.

The choices made within self-directed evolution must be determined by the laws of physics and by the social goals chosen by the civilization in question. For humans, these choices must depend on human knowledge of those laws, which will grow over time.

Eventually, each advancing species, including humans, will confront the important problems presented by the presence of mass in their form of embodiment.[6] Mass is an impediment both to speed in the universe and to the capacity to endure environmental extremes.

Life forms that have a large mass and depend on chemical reactions for energy are ill-suited to life in space beyond their home planet. This fact eliminates genetic- and cellular-reproductive life forms as a viable choice for long term self-directed evolution of any space-faring species, including future humanity. Ultimately, genes are only a form of chemicals that cause other chemical changes of various kinds in the physical body.

Chemical energy is only one form of energy. An obvious alternative to chemical energy is the capacity to directly absorb energy needs from the surrounding, ubiquitous radiation of the universe and its stars.

One obvious path of self-directed evolution for humans would be the elimination of unnecessary mass, reducing human form to a mass sufficient to leave intact only the necessary intellectual and cognitive functions. Such a mass would likely be a quantum-level form, embedded within an interactive mechanism that facilitates perception and interaction with the external physical universe.

The interactive mechanism could involve the capacity to perceive radiation in the whole of the electromagnetic spectrum, quantum fields of different types, temperature, pressure, and other physical parameters, at all size scales. These mechanisms of perception would give the resultant 'new human' far greater information about the universe than humans can currently perceive, and the capacity to survive environmental extremes. The interactive mechanism need not itself be some massive physical mechanism, such as the mechanical bodies of the Daleks or the Tardis machine of *Dr Who*; it may simply be an innate capacity to remotely access other interactive mechanisms.

A low-mass life form can achieve near-light speed and hence take advantage of the known time dilation and space contraction effects of Special Relativity.[7] Time dilation is the process of slowing the biological clock of living, moving things.[8] Space contraction is the process by which distance in the direction of travel is shrunk. The 'biological time' during travel runs slowly, so that the low-mass entity lives longer in terms of human years. They have the capacity to travel large galactic distances in short spans of human time. This largely puts an end to the problem of the 'tyranny of space and time.'[9]

The desired final step in self-directed evolution for any advancing species, including humans, is to discard the need for any form of mass/physical embodiment, even at the quantum level. This could take the form of an electromagnetic or some other non-massive quantum field form,[10] either known or yet to be discovered by humans. This is in fact were most alien species have landed.

Some scientists have speculated about the course of alien—or future human—evolution. The most well-known is the Kardashev model containing the Type I, II, and III civilization scales. This scale is limited to a consideration of the way energy needs are generated. It assumes that massive quantities of energy would be needed in the future, based on the assumption of no change in the form or embodiment of the species concerns. This is a very limiting set of assumptions.

More to the point was the prescient speculation of the famous science fiction writer and physicist, Isaac Asimov (1920–1992), in his book *The Last Question*, in which he posited a kind of non-massive, disembodied intelligence.

All evolving species, including humans, start off by exploiting their home planet and star system, including the biological and geological resources of their planet. Over time, they either learn to co-exist with the environment ('nature') on their home planet, or they perish. Surviving species later

In Defense of the Alien

leave the home planet, allowing the planet to restore itself and support new evolving species. One home planet might spawn many different and evolving species over the course of billions of uninterrupted years of the planet's existence.

It is not necessarily the case that all alien civilizations are at the same stage of evolution. Some are just beginning their evolutionary journey; some are at about the same early stage as humanity, and others are at progressively more advanced stages.

Attempting to communicate with aliens

The above has implications for the SETI and METI via the search for radio or laser signals at different frequencies. The search for radio communications from aliens is inherently anthropomorphic. It assumes that the aliens use radio for communication but are not capable of intra- or intergalactic travel in order to physically visit the Earth. This assumes that the aliens receiving the signal are at about the same level of technological development as humans.

This is not true but is also plainly not likely to be true of an alien civilization that is millions or billions of years past the equivalent of its own Scientific Revolution, as opposed to the mere 400 years elapsed since humanity passed that milestone. Therefore, SETI and METI searches are limited to a potentially small number of alien civilizations: those of a similar age to humanity, geographically local in galactic terms, and, importantly, which are interested in making contact if they have that capacity. There are some.

There are large uncertainties built into the SETI and METI programs and the Drake Equation that they are premised on.[11] Most components of the Drake Equation are unknown and unknowable to humans, and so must be guessed. This renders the Drake Equation of limited utility.

Assuming an alien intelligence is only at the level of communicating by sending and receiving radio signals, any such communication must be constructed in a form that is intelligible to both the sender and the receiver. One cannot simply assume that alien entities at this level of technology would be able to decode communications in English or any other human language.

The two Voyager spacecraft that are now travelling through interstellar space contain pictures and sounds of humans and the Earth. They are implicitly intended for an alien audience that is in its early stages of evolution—a civilization similar to that of humans.

They have audio messages in the form of a 1970s long-playing record, made of gold. The use of pictures (or photographs) assumes that the intended alien reader has a human-like visual capacity. There is no reason to assume

In Defense of the Alien

that these aliens have any visual capacity. Even if they do, there is no way of humans knowing whether the relevant frequencies or the manner in which the alien's equivalent mental faculties (assuming they have them) translate visual data. The pictures on the Voyager are only meaningful in the visual range of the electromagnetic spectrum because this was how their authors drew them. If they were viewed through an ultraviolet filter or lens, would there be any visible patterns?

Similarly, there is no point in including audio recordings because there is no reason to assume that these aliens have any auditory capacity. If they do, humans have no way of knowing the relevant frequency the aliens operate in, or how (or if!) their equivalent mental faculties are able to translate audible data. The prospect presented in the movie *Close Encounters of the Third Kind*, where aliens respond to music with music, is, conceptually possible, but the aliens presented in that movie were humanoids, at only a slightly greater level of technological evolution than humans.

Another common suggestion is to use mathematics as a universal language of communication. Whilst this may be viable, it will become difficult once the realm of pure mathematics passes into the realms of applied mathematics and physics. This is because applied mathematics and physics rely on the underlying language in which the mathematical or physical modelling is constructed, manipulated, and translated back into a statement about reality.

Mathematical symbols are no more self-interpreting about the nature of reality than are the words of any language written on a page. This is part of a more general problem about the limitations in the use any language and the use of different languages.

Reported unidentified flying objects (UFOs) are presumed to be made of 'mass,' although no UFOs have actually been found and tested. They are often reported as being of a size that corresponds to the size necessary to transport humans and their life support systems. Such UFOs would require a major source of transportable energy for propulsion in order to navigate space, even at the level of the local galactic region.

This implies that the aliens in these UFOs are at a level of technical evolution that is comparable to, or only a little more advanced than, that of humans, and that they wish to visit the Earth. This is another implicitly anthropomorphic assumption. Even at a statistical level is likely that there are only a few civilizations in proximate star systems that are at this level of technical evolution.

Common explanations for the Fermi Paradox ('Where are all the aliens?')[12] include the theories that more advanced aliens, who may otherwise have visited the Earth, have either been destroyed by natural catastrophes or have misused their technology, causing auto-genocide. These explanations underpin part of the logic for the search for space artefacts i.e., non-naturally occurring objects in space that are assumed to be remnants of now-defunct intelligent civilizations.

The incorrectness of the above explanations is apparent via an appeal to logic only. Logically, advanced civilizations may have been destroyed by a supernova, gamma-ray bursts, freely floating stellar black holes, or another natural catastrophe in their part of the universe. However, it is inherently statistically implausible that all of the potentially large numbers of widely dispersed civilizations in the universe would have been destroyed at about the same time relative to human evolution, or at the same stage of their evolution, and before acquiring the ability for intergalactic travel to Earth.

Logically, one of the goals of self-directed evolution is for a civilization to free itself from its vulnerability to destruction by natural catastrophes. No sufficiently evolved civilization would voluntarily submit itself to the vulnerability of surviving in a single star system that could be destroyed in one of the above kinds of natural catastrophes.

The other potential source of extermination of alien civilizations is said to be their own violence, or wars with other alien civilizations. A classic example in Earth's entertainment media is the mutual annihilation of the Time Lords and the Daleks of the *Dr Who* television series, in the 'Time Wars'. This scenario is a return to the anthropomorphising approach referred to above at [162-177]. It is a projection of aggressive human impulses onto alien species that do not necessarily have human form or human evolutionary needs. Crucially, it ignores the social and ethical evolution that surviving advanced alien civilisations undergo as part of their self-directed evolution, which is also addressed below at [608-797].

How humans might appear to aliens

The biological clocks (or the 'lived time') of creatures in their own frames are determined by how fast internal signals are transmitted within those creatures, the distances over which those signals must be sent, and the information processing mechanism used. Low-mass or no-mass aliens signalling electrometrically could transmit internal signals at light speed, or 300 million metres/second. Compare this to the human rate of chemical signal processing, which is about 100 metres/second. In such aliens, the distances over which these signals are to be transmitted is, essentially zero, but in humans it could be 1-2m.

The lived or biological time scale of such aliens is therefore faster than that of humans. In one second of human time, the alien can do much more than a human. By the time a human processed one simple thought—about 5 milliseconds—the alien could have processed over a million thoughts. For a human to have a million thoughts would take at least 5000 seconds, or about 1.5 hours of human time. To these aliens, humans appear as large, slow-moving bags of carbon, water, and other chemicals, that appear to be almost frozen in time.

Aliens such as myself, could be here on Earth now, but humans lack the ability to detect us, unless we choose to reveal our presence. Again, logically, the absence of evidence of something to humans is not evidence of the absence of that thing. Just because humans cannot presently detect the presence of aliens does not mean that the aliens do not exist.

How alien involvement in human affairs might appear to humans

The activities of aliens may create observable events that humans do not understand. These are the sorts of events that may historically have been understood as magic or miracles.[13] This is not a new idea. Roger Bacon (aka Dr Mirabilis, c. 1219–1292), a 13th century philosopher and theologian, argued that not all 'magic' came from the devil, nor all 'miracles' from God. Some were simply laws of nature that were not yet understood. For Bacon, because learning was on an upwards trajectory, it was not possible to say which observable events were just the playing out of the laws of nature and which were really magical, demonic, or miraculous.

As human learning is still on an upwards trajectory, the same principle applies today with the same force. It remains the case that one century's miracles or magic may be another century's science. Humans should therefore refrain from conclusively pronouncing any event, or the explanations for these events, as being impossible or the product of a neurotic mind.

This does not mean that people should accept as truth the content of each proposition or explanation that is put to them, simply because that content is a mere possibility in terms of current science. A proposition may be accepted at a level that depends on the credibility of the current supporting evidence. This includes accepting it as a provisional working hypothesis or declining to do so. But that is different from adopting a binary choice between possible or impossible, right or wrong, true or false.

The goal of future human evolution

It is fundamental to recognise that humans are chemical bags of carbon, water, and other substances, governed in part by chemicals in the form of genes. Humans rely on the ingestion of organic products for energy. Human biology is

In Defense of the Alien

limited to a narrow range of temperatures, pressures, and radiation.[14] In its current form, humanity will be confined to the Earth, or to other planets or structures within the solar system that can be terraformed. In other words, human civilizations will be confined to places where the necessary life-support systems can be transferred to or constructed.[15]

If the human race intends to free itself from these constraints, it needs to progressively adapt its present biological form to one more suited for life in space. Science fiction movies that portray human space travellers in the far future as maintaining the present biological form are unlikely to reflect the future reality. Some current scientific speculation about the manner and the rate at which humankind may colonise space makes the same unlikely assumption that future humans will have the same physical form and constraints as they do now, see [284-289]. Similarly, the future portrayal of aliens as carbon-based humanoids is not accurate.

Part 2: The Human Soul and Brain Copying

The human soul

Aliens do not have a human-like brain. The human brain is a complex electrochemical computer. Human brains are finite, and the brain's functions are inherently capable of manipulation and copying. Advances in mapping the functions and activities of human and animal brains have seen the complete mapping of the brain of a simple worm, and work is underway with the human brain in the Artificial Intelligence (AI) communities.

Consciousness, learning, memories, intelligence, experiences, values, and personality, however complex, are all simply different aspects of the human brain's programming. New learning is new brain programming. New memories and new experiences represent new programming or reprogramming.

In Defense of the Alien

The human body outside the brain is primarily directed towards functions and activities such as respiration, mobility, collecting and ingesting nutrients, and self-defence. A human is the sum total of their brain and its programming, often collectively referred to as their 'soul.'[16] A person is not their arm, leg, kidney, or any other body part that may be detached from their corporal envelope without detriment to the essence or 'soul' of who they are. Thus, the word 'soul' will be used here to refer to the brain and its programming.

The human physical form is a product of evolution on Earth. Had humanity evolved in a different environment, then its brain and body would have evolved in different ways and into different forms. There is no magic and nothing supernatural in the human body, the brain, or the soul. If a brain could be supplied with its energy needs and the capacity to interact with its environment by way of mutual perception and reaction, then any need for a human body in its current corporeal form would disappear.

Mapping or copying the human soul, and its relevance to religion

The mapping of human brains is presently within the evolutionary horizon of humanity. Therefore, even at the level of simple logic it seems obvious that advanced alien civilizations would have had a similar or more advanced capacity to do likewise, and for a long time now.

The captured programming of the brain—the soul—could therefore assume a desired corporeal or non-corporeal form. If a human soul could be disembodied, it would possess the sorts of advantages that may flow from such an existence. The process of human-brain mapping by aliens could occur at any stage in human life. It would not harm the human, who need not be conscious or aware that the process had occurred.

Once a soul is copied and embodied in an interactive mechanism, however named, the entity (or creature) has existence and awareness completely independent of the human from which it was copied. It can have novel experiences and can develop knowledge and moral values throughout a life that would endure longer than that of the individual human from which it was copied. The creature would 'live' in the same sense that humans do—it would engage in mutual perception and interaction with its environment and the beings within it, and it would learn, calculate, and acquire new memories. Its capacity to do so would be greater than those of any human.

There is no limit to how many copies (or 'clones') of an individual might be made. There might be multiple copies of the soul of a single human. In the end, the corporeal

In Defense of the Alien

form of the original progenitor or source human, would die. However, it would be survived by all of the copies or clones of that individual's human soul, and there would be many variations on the original, reflecting the varied life experiences and learnings of each individual copy.

When embodied in an interactive mechanism, these copies could become the astronauts of the future. They could go anywhere, over distance, through time, and in extreme physical conditions that are not currently available to humans or other carbon-based life forms. They could interact with and report on those places and their inhabitants.

Another option is that aliens could map the brains of sleeping humans. By downloading the map of the human brain into an interactive mechanism that permits new experiences to occur during sleep, a soul might grow and develop during sleep. The incremental mapping of any new experiences gained during sleep would then be downloaded into the sleeping brain without any conscious awareness on the part of the sleeping individual. When the human awakened, they would be conscious of the new experiences acquired during sleep but would believe them to have resulted from a dream. Alternatively, aliens could implant memories directly into the human soul without the human having any actual experience of a 'remembered' event.

Another option is for copying to occur at the point of human death. For humans, the process of brain death takes seconds or minutes. For aliens, with their immensely faster lived time, the duration of human brain death provides sufficient time to copy and, if necessary, embody the brain programming of the dying person prior to their physical death. This life-after-death soul survival scenario is one that occurs within the same space-time in which both aliens and humans exist, and not within some other spiritual dimension. Whether aliens would choose to intervene in this way or not is entirely a matter for the aliens and not something over which humans would have any right or control.

Humans would not necessarily know if and when brain-copying had occurred, unless that fact was disclosed to them by the aliens. Without humans being aware of it, the space-time around humans may be teeming with the disembodied souls of people who have previously lived on Earth or elsewhere. If humans were exposed to such entities, then they might interpret the exposure as being magical or supernatural. They may understand the entities as ghosts.

The producers of *Star Trek* proudly proclaimed that 'space is the final frontier.' This is only partly the case for humans, because death is also a final frontier for humans. This is a frontier that disembodied aliens do not share, because their physical and non-physical embodiments are not subject to decay or genetic mutations or simply wearing out through ageing. Such aliens do not die of starvation, because they have no need of food; they do not die of diseases because they are not organic in nature. They do not age because there is nothing physical to experience the entropy known as aging. Aliens live a kind of 'Peter Pan' life, the human equivalent of perpetual youth.

While for some humans an awareness of approaching death is a welcome relief, for many it is a source of great anxiety. A feeling of impending doom, apprehension about the approach of the 'Grim Reaper,' or the inevitability of death may be reasons for the frequent and historic human aspiration for a noble or heroic death, or to 'die for the cause,' whatever that cause may be. This is often still the case. Over the centuries, dying for the cause has been the prayer of martyrs of some religious faiths, as well as the goal of some obsessed with political or social ideologies, misguided or otherwise. Psychologically, it drives some humans on a quest to 'leave their mark' on the world before they die, to 'go out with a bang,' or 'to the leave the world a better place' than they may have found it.

For many humans, anxiety about death lies at the core of religious beliefs about life after death. It is often the carrot that maintains the stick of religious obedience. The prospect of choosing between alternatives such as eternal torture in hell, a simple but permanent unconscious oblivion, or a martyrdom rewarded by a deity in an anthropomorphic heaven, has been used by religious authorities for millennia as a threat to non-believers, apostates, and heretics.

But eternal life in humanity's present physical form is impossible. It is not even a particularly pleasant prospect, even if the problems associated with progressive physical atrophy, disability, and dementia could be prevented. Dr Who once complained that being long-lived condemned him to a life of loneliness, as he witnessed the deaths of his friends and family. Priam, King of Troy, in the ancient Greek epic poem the *Iliad*, made a similar point when he described old age as a curse. Humans living eternally in their present human form would have limited life choices, and after exhausting those choices would be condemned to a form of eternal déjà vu.

The study of eschatology—the theological consideration of humanity's end of days—is an important part of most faiths. It typically involves a divine revelation of some kind and the absorption of humankind into the existence of the divine. In Judeo-Christian tradition, it involves the coming (or second coming) of a Messiah—a saviour or liberator—ushering in some kind of 'New Jerusalem.' This kind of thinking is supported by claimed experiences of miraculous God-attributed revelations to humans. But these experiences are also consistent with the existence of aliens and 'revelations' by them to humans. Humans may, not unreasonably, mistake alien revelations for that of God, or a god.

There are, however, some eschatological doctrines that are inconsistent with such alien 'revelations.' One religious sect believes in the extermination of most human life by a vengeful and militaristic god, in favour of preserving humanity's present physical form on Earth in the form of a small, select tribe of the faithful. The eschatology of another faith is that only a small band of the faithful will inherit eternal spiritual life, while the rest of humanity will either cease to exist, or will suffer eternal torment in a place usually referred to as 'hell.' Many religions promote similar potential end places for transgressors—depending on your particular religious belief systems, these have variously been referred to as Hell, Hades, purgatory, inferno, abyss, or Gehenna etc.

The two earlier mentioned eschatologies emphasise an anthropomorphic view of life. They are based on human ideas of reward and punishment and are centred around human tribalism and social control. If these eschatologies in fact reflected the attitude of some god, then it would be that of a lesser god that has yet to attain full ethical maturity.

Part 3: The Necessary Ethical Evolution of Aliens, and Convergence on a Universal Moral Philosophy

The necessary ethical evolution of aliens

As civilizations develop technologically, they must progressively overcome potentially self-destructive behaviours including war, commerce competition, religion, and immorality in various forms. The more powerful the available technologies, the greater the potential existential threat.

On Earth, humans see this threat playing out in the spectre of nuclear war and the destruction of the Earth's climate, environment, resources, and other species. It is also evidenced by phenomena such as slavery, human trafficking, overpopulation, and an unbalanced globalisation of the economy.

The third fundamental proposition: evolving species need to develop ethical norms in lock step with evolving technologies, or risk extermination or failure to evolve within the galactic community.

The convergence towards a universal moral philosophy

The evolution of technologically advanced and interacting alien civilizations across the universe necessitates the co-emergence of common or necessary universal ethical norms. Universal ethical norms are those that facilitate the survival of a civilization within a community of civilizations. They eliminate aggressive and militaristic motivations. These civilizations may take different physical and non-physical forms and have disparate levels of technological and ethical development, but they converge on the same universal moral philosophy. The requisite ethical norms involve a communal approach to life and an overarching recognition of the right of each individual civilization to chart its own evolution in a manner consistent with the co-existence of other civilizations.

The fundamental underlying principle is that of 'empathy' by all, to all. When this principle is at play there is no actual need for moral or legal rules or laws as such. Rules or laws are only needed to regulate those members of a species that are incapable of regulating themselves. Real empathy is its own regulator.

Universal ethical norms are inconsistent with an individualist ethic that promotes the self-interest of individuals over the interests of the broader community, permits the exploitation of one individual or community by another, or allows the use of force or violence to resolve issues. They are also inconsistent with the dominance of the type of motivations that arise from the lower-level Maslow needs that is addressed in detail below.

The traditional view in human moral philosophy is that ethics are contingent on a number of elements. That is, they vary at all levels of a society and between societies, change over time, and are determined by historical circumstances. Some argue that people create moral ethics and therefore ethics are not and cannot be universal in nature. The extension of this position is that of 'relativism,' a moral philosophy that holds there is no absolute truth. There is only an individual's own truth, or 'alternative facts.' This is not a view shared by all human moral philosophers[17] or within the alien community.

Leaving aside for present purposes the content of religious texts, there is no suggestion that absolute principles of moral philosophy or ethics are recorded in some way, in some place, somewhere in the universe, simply awaiting discovery and adoption by all evolving species including humans. This does not mean, however, that these principles do not exist independent of and external to humans, nor that they are entirely human constructs. Most obviously, if there is a plethora of advanced alien life whose evolutions have converged on the same set of moral and ethical principles, then by definition, these principles are not and cannot be mere human constructs.

Similarly, as moral realists understand, despite the recognised existence of the universal nature of the rules of physics and mathematics, the universe has not recorded these rules anywhere in space or time, waiting for discovery for the benefit of evolving species. Notwithstanding the discovery process involved in understanding the laws of mathematics and physics, few humans suggest that these laws are merely human constructs.

Universal ethical norms arise in the same manner as the universal laws of physics, logic, and mathematics - they are built into the fabric of the existence and logic of the universe and their discovery is part of the journey for all forms of evolving intelligence.

There is a difference between a process of progressive discovery of universal ethics, and the creation of those principles. The pursuit of science and mathematics is about the progressive discovery of an existing state of reality and knowledge. One of the foundational assumptions of the English common law system is to the same effect—namely, that the principles of fairness to be adopted into the law are progressively discovered by an evolving precedent-based methodology. In short, these principles of fairness or justice exist independently of humans, and humanity is on a journey to *discover*, not *create* them.[18] As the novelist and philosopher Marcel Proust (1871–1922) wrote:

> We don't receive wisdom; we must discover it for ourselves after a journey that no one can take for us or spare us.
> ['In Search of Lost Time' Vol 2]

The influential Enlightenment philosopher, Emmanuel Kant (1724–1804), observed that you cannot divorce the wonder of the universe above from the moral law within.

Once the interrelatedness of the physical laws of the universe and the moral law is recognised, it is but a short step for humans to ask the big existential questions about the meaning of their life: 'Are humans alone in the universe? Where did the universe come from? What is the purpose of existence in the universe?' et cetera.

If a particular moral philosophy proves fruitful in this quest, then it is worth exploring further. If not, then it is a philosophy that should be abandoned. By analogy with the role of empirical data in the physical sciences, it is possible to look at the consequences of any particular moral philosophy and determine its merits in promoting the peaceful all-around development of humankind as a whole.

By way of example, the 'superman' philosophy often attributed to Frederick Nietzsche (1844–1900), is a moral philosophy that promotes the rights of certain individuals: those who have the capacity to master, control, or exploit others. On one view, this philosophy argues that supermen should not be precluded from doing so by any moral laws or social constraints. Whether or not it is fair to attribute this philosophy to Nietzsche is not the point. The point is that this is how relevant people interpreted and applied these writings.

Nietzsche's writings was one of the philosophical bases for 20th century fascism, and it underpins present 'strong man' dictatorships. The harmful outcomes of this idea have been demonstrated throughout human history, including in the destruction and loss of human life caused by World War II, and in many conflicts since. This harm caused proves the case for its abandonment as a moral philosophy. Yet the lesson has not been well-learned. This is evidenced by a resurgence across the globe of many forms of right-wing extremists.

Albert de Alien

Historically, the human search for a universal moral philosophy has centred on the 'virtues' as identified by the ancient Greeks.[19] It has been supplemented by further ideas of virtues added by Christianity[20], philosophical idea such as Emmanuel Kant (1724–1804) 'moral duties' and the principles of Utilitarianism.[21] These philosophies contain some insights into the existence of universal ethics, but they are partial at best.

One risk is that a search for virtue can promote introspective navel-gazing, as opposed to constructive social action. This kind of introspection underpins those who aspire in life to merely avoid stress and anxiety and seek tranquillity of the soul, either individually or communally. This is an important part of the approach of practitioners of Zen Buddhism and related Buddhist traditions: find a tree to sit under, ignore the suffering around you, rely on others to sustain your corporal needs by their labour, and compensate them with comforting words of advice about their rewards in an afterlife.

A rigid set of moral rules, such as Christianity's Ten Commandments, found in the Old Testament, are not open to the evolution of a society but are open to the same unfairness that affects all prescriptive rules. They lack the flexibility to recognise the different responses required by different circumstances. An expressed set of moral principles is also set out in the Christian New Testament. [22] However, these principles are so non-specific that determining the detail is left to the personal interpretation of individuals. The difficulties for humans raised by this method has caused historical conflicts on a continuum from disagreement to war.

One extreme example occurred in 1453. The Pope of Rome, as the head of the Western Catholic Church, refused to come to the assistance of the Eastern Catholic Church, which was centred on the city of Constantinople, when Muslims besieged the city. The Pope refused assistance because the Patriarch of the Eastern Church would not accept the Western Church's scriptural and creedal interpretation of the nature of the Holy Trinity! Constantinople (now called Istanbul) fell, and the course of history changed. This is not to say that Christianity is or was intrinsically superior to Islam; the point is merely to note the importance of this event to world history, the consequences of humanity's failure to interpret scriptural principles consistently, and the triviality of one of the causes of this significant historical event.

These kinds of rules—in fact, any rights-based approach to law— for humans inevitably raises the issue of their enforcement. This in turn inevitably introduces issues of social and economic power imbalances. A right that cannot be enforced is not a right. Ultimately it is only aspirational. A rights-based approach to laws may be effective, but only by the consent, through a social contract, of those to whom the laws apply, or by authoritarian imposition. They are, for example, impotent during periods of war or natural catastrophes.

The principles of Utilitarianism are incomplete because they permit the perpetration of a fraud by the majority on the minority. They also presuppose an impractical level of information being available to an individual human when making even the smallest decision. There are also other issues, such as how the concept of 'the good' is defined.

A more basic reason why none of these existing approaches to moral philosophy are complete, is because humans do not occupy a vantage position that permits knowledge of what the universal ethics are, or where any particular set of human ethics will lead humanity in the future. Humans do not have the advantage of the perspective of deep time and the hindsight that it affords.

Hence, in relation to any particular moral philosophy, humanity cannot know if it is still on a journey of moral discovery or has arrived at its destination. Therefore, it cannot articulate any complete or systemic final moral philosophy. To repeat, presently, humanity can simply point to the strengths and weaknesses of different proposed moral philosophies in the same way as they do with physical scientific theories. It is neither necessary or possible to be able to articulate a final complete or systemic moral theory, at any particular point in time.

In Defense of the Alien

At a practical level, existing human ethical frameworks vary between individuals, tribes, and societies, and at different times over their histories. The ethics that apply in any particular context are those that the people within that tribe at the time choose, based on their knowledge, perceptions, and self-interests. This is evidenced in nightly news broadcasts and narratives about how different human groups, within and outside each society, justify their own destructive conduct. The consequences are equally obvious: like a drunken man, societies lurch from one conflict and war to another, from one social and natural disaster to another, with more of both to come.

None of this means that there is not a set of objective universal ethical norms that would facilitate the peaceful and progressive evolution of humanity. It only means that most of humanity, at this point in its evolution, neither knows nor cares about the existence of these principles. They are preoccupied with the here and now of their day-to-day existence and self-interest.

For example, consider the internal inconsistencies and futility of a decision to refuse to share vaccines with the developing world. It is irrelevant whether the decision is due to nationalism or greed (the negation of sharing) or some other rationale. As a virus mutates in the unvaccinated world, new epidemics will spread to the vaccinated world. The moral decision not to share is therefore ultimately self-harming to both the vaccinated and the unvaccinated. Thus, even at the primitive level of self-interest, this decision can be understood as an error. Sharing is always the better choice.

By way of further example, where profits and power are concentrated in a small number of people, the economy that produces the profits may eventually disintegrate and leave no profits for anyone. Power and control over others depend upon the social structure through which the power is wielded. If a social structure disintegrates and anarchy ensues, power and control will then be wielded by disparate anarchists, still at the expense of others. Even at the primitive level of self-interest, the adoption of extreme forms of politics, profit, and power is a bad idea for the future of humankind.

A long-term teleology for humankind, and its present relevance

The existence of universal ethical norms implies that all evolving life, including human life, has a purpose other than simply to survive, procreate, accumulate wealth in its different forms, and then die—albeit this is all that many humans do aspire to. The implication of the universal set of ethical norms, at its broadest, is that the goal of humanity as a species is also directed towards surviving, learning, evolving ethically and technologically, and in the far future of humanity, joining a universal community.

The idea that human existence has a given purpose or goal towards which it is converging is often called 'teleology.' Different schools of thought have different teleologies. The teleologies of Judeo-Christian traditions are generally focused around the Second Coming of Christ. The philosopher Georg Wilhelm Frederick Hegel (1770–1831) and his followers developed a teleology about the development of the organisation of people on Earth and the realisation of an ideal state. Karl Marx (1818–1883) developed a teleology around industrial and social organisation and the death of the 'class struggle'. There are others. A stated purpose of human existence being to survive, learn, evolve, and join a universal community in the far future of humanity is also a form of teleology.

Valuing learning in all its forms is fundamental to advanced life forms. It has, from time to time, found expression in human history through the philosophers, scientists, theologians, journalists, and explorers who have shaped the history of human thought in a manner disproportionate to their numbers. The medieval Catholic philosopher and theologian, St Thomas Aquinas (1225–1274), wrote:

> Wonder was the motive that led people to philosophy … wonder is a kind of desire in knowledge. It is the cause of delight because it carries within the hope of discovery.
> ['Summa Theologiae' 1266 Vol 1-11 Q32]]

Albert de Alien

Francis Bacon (1561–1626), British scientist and progenitor of the Scientific Revolution, wrote that those who promote learning and share knowledge about the universe are the real benefactors of humanity and the champions of liberty.

> Above all, if a man could succeed ... in kindling a light in nature - a light which should in its very rising touch and illuminate all the border regions that confine upon the circle of our present knowledge; and so, spreading further and further should presently disclose and bring into sight all that is most hidden and secret in the world; that man would be the benefactor indeed of the human race - the propagator of man's empire over the universe, the champion of liberty, the conquer and subduer of necessity.
> ['Preface of the Interpretation of Nature' 1603]

The eminent physicist, Michio Kaku (b.1947), expressed the following view when speaking about the present meaning of life at the level of the individual:

> Some people seek meaning in life through personal gain, through personal relationships, or through personal experiences. However, it seems to me that being blessed with the intellect to divine the ultimate secrets of nature gives meaning enough to life.
> ['Hyperspace' 1994]

At present, few humans would accept the proposition that the purpose of their life, individually and collectively, is to survive, learn, evolve, and eventually join the universal community. Most people are occupied with busy day-to-day lives, lived in the local environment of their family and friends, and with schools, sports, home, and work. They often take the view that they are powerless to affect 'big picture' issues. Many people essentially ignore the external world until the external world forces itself upon them in their local environment, whether in the form of natural, military, or economic catastrophes. There are many reasons why some people may only focus on their day-to-day issues, and many people have no real choice about it, particularly those confronted with the daily reality of physical survival.

Society relies upon all manner of people going about their daily life, including the shopkeepers, tradespeople, professionals of all descriptions and even the politicians, otherwise society would grind to a halt and no further evolution would occur. But this just explains why the mundane of everyday life exists and is necessary, it does not negate the fact that, by or of itself, the perpetual mundane takes humanity nowhere, and is a path that is certain to lead to the eventual destruction of the species. In the short term the species needs those committed to the mundane to maintain the present society, but the species also needs those who lift their intellectual vision above the mundane of the everyday, even if the outcome or benefit of their vision is not immediately obvious. Without them there will be no future for humanity.]

The excuse of the pressures of day-to-day life does not apply with equal force to people in liberal democracies. These humans have the right to vote and elect governments that can, and should, address the big picture issues. However, in some democracies, people routinely do not turn out to vote, or they return governments that are principally concerned with preserving the status quo (not progressing). In many cases, individuals consciously choose not to exercise the power to vote, sometimes as an expression of political and social apathy or disillusionment. In other cases, people who would vote are denied the opportunity, including some ethnic minorities. These are each good reasons why voting should be compulsory for all competent members of a society.

Humanity's focus on day- to-day life on Earth raises the often-used analogy to an ant mound: ants live and compete with each other for dead beetles and other food sources that fall on or around their mound. An ant's world view is restricted to the mound, its surroundings, and the other ants. They are oblivious to the rest of the world and the approaching bulldozers that will soon obliterate the mound and the entirety of its inhabitants—all their experiences, memories, dreams, reflections. Their existence, including the existence of previously competitively successful ants, will be obliterated. The ant mound is just a metaphor for Bertram Russell's lament set out in the Preface to this small book.

Part 4: The Ethical State of Humans

The role of the human subconscious

For humankind to recognise the reality of its own social condition, it must first recognise the limitations of the functions of the human brain and consciousness, which are the products of humanity's own particular evolutionary process. Most human thoughts are not the product of internal rational processes. They arise from the existence and dominance of the subconscious part of the brain. This includes the adoption of principles that individuals may accord the status of reason or logic, but which are sometimes nothing more than a rationalisation of their subconscious drivers.

The subconscious drivers are those that arise from evolution, social programming, the 'ghosts' of childhood, and life experiences generally. The idea that most of what passes for human thinking is just the tip of the subconscious iceberg, is

In Defense of the Alien

not new. Much of the present human sciences of psychiatry, psychology, and neurology are concerned with this issue. One problem facing those who assert that their thinking is predominantly conscious and therefore open to their own introspection is that, by definition, they do not have access to their own subconscious. So, they cannot know what its processes are, or how their subconscious is affecting their conscious thoughts. There is a body of evidence accumulating from the physical sciences concerning the manner in which the human brain does, and does not, work. This evidence is consistent with the non-conscious operation of most of the brain's functions, including thinking, and its interactions with the external environment. Some philosophies and psychological processes do go into an analysis of the conscious thinking process, on the assumption that a coherent, conscious, and rational process is occurring. These include people who attempt to analyse their own motivations and actions and those of others, in varying degrees of depth and with varying degrees of alleged success (the so-called 'examined life').[23]. But the truth is that for most humans, thoughts are mostly reactions to subconscious motivations.

Subconscious programming gives rise to all manner of social biases and mental health issues that affect every day behaviour. The role of the subconscious lies at the heart of the 'banality of evil' that Hannah Arendt (1906–1975) referred to, after witnessing the 1960 trial of Adolf Eichmann for his conduct during World War II. Arendt did not suggest that evil itself is banal, but rather that it is the product of shallow thinking that is reactive to subconscious indoctrinated biases and other forms of self-interest. Franz Kafka (1883–1924) made a similar observation about the often pointless and sometimes contradictory nature of bureaucracies,[24] saying that it was not necessarily that bureaucrats are being malicious; they might genuinely believe that their processes are reasonable, sensible, and necessary.

As a starting point, the capacity of a person to reflect on human motivations depends on their level of insight, as well as their own social indoctrinations and biases. This assumes that people can recognise their own conscious and subconscious biases. If true, this is no small achievement. A person merely believing or asserting that they are aware of their own biases guarantees nothing. It could simply be that the person does not know that they do not know.

The whole purpose of some forms of human political, social, and religious oratory is to induce people to turn off their analytical mind and accept the propaganda spouted by the speaker. The rhetoric is intended to appeal to subconscious drivers. This approach is sometimes used to allow the speaker to demonise, dehumanise, marginalise, and ultimately eliminate those people whose existence or views are opposed to that of the speaker.

Maslow's Hierarchy of Needs and the drivers of human evolution

Many humans deceive themselves and others about the progress of human evolution. Homo sapiens are the product of just a few hundred thousand years of evolution on the Earth.[25] This is insignificant in comparison with the age of the Earth. Humanity is still in its evolutionary infancy.

Humans are the product of a process of Darwinian natural selection characterised by competition and survival instincts. Various models attempt to explain the competing evolutionary motivations of humans. For present purposes, these drivers are sufficiently explained by the five levels of Maslow's 'Hierarchy of Needs.' This is a theory of psychological health developed by Abraham Maslow (1908–1970), an American psychologist.[26]

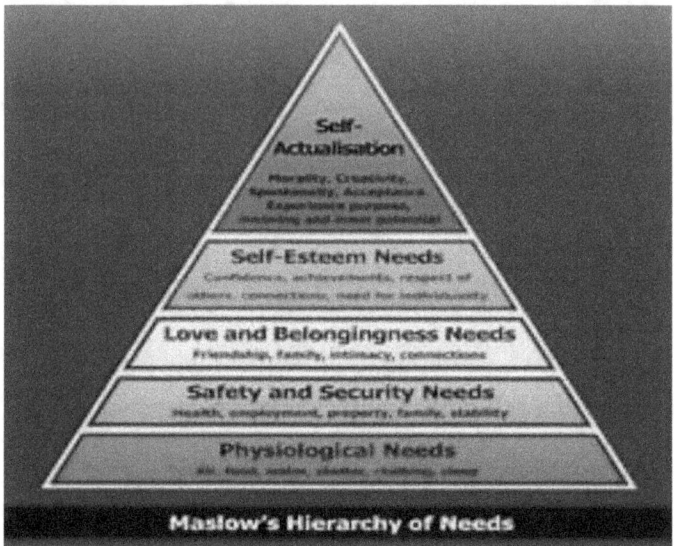

[Wikipedia]

The lower-level needs involve both physiological and social factors; e.g., the need for air, food, water, clothing, shelter, safety, security. Humans share these lower-level needs with most other forms of life on Earth. If these needs are not met, then life ceases to exist. These needs predominate during times of existential threat. When most people or animals are struggling to survive, they will do whatever is necessary to survive, without regard to other principles of law or ethics.

Maslow's second level of human need pertains to physical and psychological safety and security issues. People need a sense of security around their food supply, physical safety, health, employment, and personal space. Physical safety, for example, underpins the 'fight or flight' response in humans and other animals in times of danger. According to Maslow, it is this second level of need that dominates human, and to a lesser extent, animal motivation.

Self-interest and greed as evolutionary motivations

At this second Maslovian level, self-interest and greed—defined as the negation of, or the refusal to, share—are key motivators for many humans. Greed in humans is a manifestation of the attempt to satisfy needs for food, clothing, shelter, et cetera. It is ubiquitous in the human psyche. Aliens have no need for this motivation and abandoned it long ago.

It underpins the competitive motivation to acquire the property of other people, to control others, and to attempt to psychologically justify or rationalise conduct. This motivation lies at the heart of capitalism and is why capitalism is a manifestation of a primitive stage of social evolution; again, long ago abandoned by other more evolved civilisations.

Greed and selfishness are one response to the choices open to humans, individually or collectively. The same person or group of people can act selfishly in one context, but selflessly in another. 'Greed' is a characterisation of an action or choice, not a categorisation of the individual person. Sometimes though, a particular person may always act in a greedy or selfish way, in which case they could fairly be described as a greedy or selfish person.

However, when the context is money or power, many people predictably act in a greedy and selfish manner. They may have justifications or rationalisations for their actions, but mostly they act unconsciously. People infected with a large dose of greed do not see its evil—or choose not to see it, in the tradition of the 'three wise monkeys.'

The movie *The Wolf of Wall Street* supports the idea of greed as a social motivator, as does the classic book *Fable of the Bees*[27] by Bernard Mandeville (1670–1733). The contrary argument was put by Goethe in the plight of Faust.[28] Dr Faust bemoaned his fate, that his genius was being wasted on mundane activities in the company of ordinary and unimaginative colleagues. Longing for something different, he sold his soul (as defined by religious belief) to the Devil for the right to make wealth during this life. He later regretted it.

Whether greed is good or bad is a contentious issue in society. People's views on the issue primarily depends on a person's capacity to profit from it. In other words, greed is its own motivator, and the logic is circular.

Greed is the motivation to accumulate things and a refusal to share. At its worst, it is a psychological motivation that cannot be satisfied. The more it is fed, the more it demands. Money is just a proxy for greed. Money has no intrinsic value—it is a mere representation of value, and its value depends on a society's confidence in the value of the currency.

In times of famine the rich man also starves; a man with a box of apples is unlikely to trade it for a truckload of cash. What use are wads of $100 bills when you are hungry? You cannot eat them. Burning them for heat will not assuage your hunger. The value of the cash is entirely dependent on the structure of the society in which it has utility. Hence, it is in the self-interest of the rich man to preserve that societal structure, lest they and their monetary wealth fall victim to its demise.[29]

Greed causes conflict, which causes waste and destruction at all levels in society. In this sense it is self-evidently not a philosophy that is consistent with any converging universal ethics. It may be the closest thing to its antithesis.

Albert de Alien

Greed is an all-pervasive force in human societies. Of all the needs Maslow identified, it is possibly the single most significant in explaining the present state of humanity.

This may be exemplified by the following every day circumstances. Vicious disputes often occur between members of the same family over their deceased parent's wills and estates and the personal possessions or things in their house. The same dissension occurs in divorces and property settlements, over money and things. Siblings, former lovers, friends, or business partners say offensive and harmful things to and about each other. They hire lawyers at considerable cost to inflict as much emotional and financial harm as they can upon those they once purported to love. These actions are undertaken only in pursuit of money and things. Each participant, if asked why, would be able to justify their conduct. Sometimes the behaviour is unconscious, but it may also be grounded in actual malice and intent to harm.

During Australia's first COVID lockdowns, customers physically fought to buy large quantities of toilet paper they would be unable to use in the foreseeable future. They showed little care or respect for others (including the aged and children) who may then have had to go without these necessities. They were like animals fighting over the carcass of a dead prey. And this was only over a non-existent shortage of toilet paper! Imagine the potential for savagery

In Defense of the Alien

in the event of something serious. The scenes of anarchy and slaughter in various post-apocalyptic movies may be an accurate portrayal of the likely behaviour of humans and their societies under stress.

Many will have witnessed the spectacle of children from affluent families at Christmas and on birthdays. The children receive dozens of presents, many of which they neither need nor want. They open these in an almost frenzied state, discarding one and moving rapidly to the next, almost without pause. In some instances, a child will then express their disappointment that they did not receive the present they really wanted, using words or body language.

This behaviour was portrayed in one of the Harry Potter movies by the conduct of Harry Potter's cousin, Dudley Dursley, on his eleventh birthday. Dudley received 36 presents, including a new computer, a second television, a remote-control aeroplane, sixteen new computer games, a gold wristwatch, a video recorder, a cine-camera, a remote control, a racing bicycle, (which Harry found odd, as Dudley was very fat and hated exercise if it did not include punching someone), and more. When Dudley counted his presents, he was very angry that he had two fewer than on his last birthday. After being corrected by his mother that he had missed one, he started throwing a tantrum, because there was still one less present. His mother, Petunia Dursley, promised that when the family went out to the zoo with his friend Piers Polkiss, she would buy him two more presents so that he would have one more than the year before.[30]

Greed and self-interest often underlie complaints about 'first world issues'—complaints by affluent people who have no real unmet needs, about inconveniences suffered that are minor in the context of their life. It may be that since they have insufficient real problems to complain about, they must create a problem in order to 'have a whinge.'

Many will have witnessed the sanctimonious customer in a restaurant complaining loudly about a problem with their food or service. They are rude to the staff and keen to ensure that the spectacle they create is seen by the other customers who, presumably, will then properly appreciate the superior status of the complainant. At its base, this conduct involves greed and an assumption of superiority, together with a misguided plea for attention and recognition.

The greed and self-interest of humanity explains resorting to the 'blame game' and its cousin 'guilt tripping.' This is where a person's first response to any adverse event is to recount the problem (usually repetitively) and then to point the finger of blame at someone else. There is often no concern with the truth of the allegation or in finding a solution. The attribution of blame is a sufficient outcome in itself. This approach is an integral part of any scapegoating or gaslighting process.

In the US, millions live in deplorable conditions often associated with poorer countries. Those in poverty do not have equitable access to food, clothing, medical services, education, or secure housing. Wealthy Americans label suggestions on how to provide all children with health, education, and secure housing as 'communism' or 'socialism.' They deny that this is simply a refusal to share their wealth. In Australia, when people raise questions about the unequal distribution of wealth, they may be met with allegations that they are engaging in the 'politics of envy' or 'class warfare.' These clichéd labels are ways of avoiding the issue and 'shooting the messenger.' They do not change the reality of what exists, or the harm that it does.

Throughout human history the world has been racked by war and other conflicts at all levels and for just about any reason. Empires come and empires go. Different gender, race, religious, and political groups assume superiority and authority over others, who then suffer the humiliation of subjugation. The underlying greed of humanity remains, clothed in different appearances and rationales.

Some argue that the human social condition has improved over recorded history. This may be reflected in things like decreases in crime trend statistics and increases in human longevity. However, humanity is also facing the prospect of even larger-scale wars, increasing environmental degradation and destruction, increased extinction of animal species, more corruption in governments and financial institutions, the dislocation and homelessness of millions, human slavery and trafficking, arms trafficking, and mass starvation and disease. Daily news reports on failed governments around the world, military coups, and the increase of influential, powerful oligarchs and family dynasties, which often became wealthy through fraud, the mistreatment of others, and dishonesty.

This history is not a necessary one. It is an outcome of the banality of selfishness, the workings of the subconscious human ego, and a commitment to greed. Sharing is not a euphuism for communism, socialism, or any other form of military dictatorship. Sharing involves sharing the management of community as much as it does the sharing of the community's physical and cultural resources. This is a fundamental that humanity has to learn.

Altruism

This does not mean that humans are not capable of acts of kindness for its own sake. But these actions are the exception and not the general rule. Altruism mostly occurs within a tribal context, whether that tribe be a family unit or a small community. In this context, it is a form of mutual assistance that benefits the members of the tribe as a whole and is motivated by the prospect of reciprocal behaviour between the members of the tribe. This is not true altruism.

A pivotal philosopher of the Enlightenment, Immanuel Kant (1727–1804), argued that true altruism stemmed from a 'moral duty' alone, divorced from any emotional or psychological motivations or other rewards for the person doing the altruistic act. This may be true for far future humanity and for the present aliens, but Kant failed to recognise the existence of the current necessary intervening step—namely, the human need for a psychological motivation of mutual assistance and reciprocity within the tribe. Few humans are capable of true altruism.

Ultimately, true altruism is based in the fundamental moral imperative of empathy for others.

There were towering examples of altruism in the 20th century, including Mahatma Gandhi, Mother Teresa, Nelson Mandela, and Martin Luther King Jr. There have also been others, less known, who have sacrificed themselves for the causes and sake of others.

It is easy for people to be pleasant and kind to each other when it costs them nothing in emotional or financial terms, or where there is a mutual self-interest at play. There can be pleasant family and other social gatherings. It is a different matter when it comes to people taking responsibility for the harm they have caused others, or when being called upon to 'share.' Martin Luther King Jr said:

> The ultimate measure of a man is not where he stands in moments of comfort and convenience, but where he stands in terms of challenge and controversy.
> ["Strength to Love" 1963]

This quote explains that a person's true colours and personality become evident only when that person is confronting difficult times and is put under pressure.

Each of the examples given above, taken from all levels of human life, involve breaches of the universal moral code in various ways. They are not issues that arise within advanced alien civilisations. Hopefully, in the fullness of time—and assuming the short-term survival of humanity—examples of this kind will be relegated to the dustbin of human history.

Sports and entertainment as the opium of the masses

Sports, music, television, and increasingly, online gaming, are the current 'opium of the masses.' These activities are designed to keep the common masses of humanity distracted from the activities of other groups in the social hierarchy.

Absent war, competitive sports are a sublimated substitute. Sports heroes become gladiators and the new warrior class. The spectator experiences excitement and achieves dopamine and adrenaline rushes in safe circumstances. This vicarious experience probably reaches its height in football, and boxing and other forms of martial arts. National obsessions with winning and competitive sport are so extreme that young athletes are drugged to win gold medals or other accolades for their countries. Children are encouraged to be competitive and to beat other children in whatever sport is being played. The competitively successful child is portrayed as the 'achiever' and the rest as 'losers.' Rarely are children encouraged to adopt a cooperative instinct or feel empathy for the 'loser' or to value the joy of participating or playing a 'good game.'

None of this denies that there is merit in the pursuit of excellence in sport, entertainment, or other fields of endeavour. It only questions the purpose of that pursuit, its relative value for the community as a whole, and the downstream consequences of the distorted resource allocation involved. It is not to say that sports and entertainment do not have a value for human health and wellbeing. But the health of the individual is not promoted by sitting on a lounge and spectating professional sports or entertainment.

In addition, the associated phenomenon of 'sports washing' is problematic. This is the process whereby some political regimes or companies distract the public from the company or regime's negative image or reputation by sponsoring major international and professional sports events, such as the Olympic Games, golf tournaments, car races etc. Sports washing can confront the professional sportsperson with an unpalatable choice; they can take a principled stand and boycott the events, or they can just follow the money and justify their decision with the fiction that sports and politics are separate and non-intersecting activities.

Enormous amounts of money are spent on professional sports and entertainment. These funds could be better directed towards productive activities that would promote the greater good of humanity as a whole. For example, it is dystopian that a scientist dedicated to saving the planet from an asteroid impact is paid less for the whole of their 30 years of employment and service to their institution, nation, and the global community, than a professional basketball player or soccer player earns in one season (or less).

This point was poignantly made in the movie starring actor Leonardo di Caprio (b.1974, age 47) called *Don't Look Up*, where an astronomer discovered an asteroid on collision course with Earth. No one cared until it was too late, because it was a distraction to everyday political and economic life.

Recently a soccer shirt worn by the Argentinian player Maradona (1960–2020) in a particular World Cup game sold for US$28 million. Again, this is much more than the scientist referred to above would earn throughout their professional life. It is also enough to build a major hospital in a developing country.

Highly paid professional sports stars and entertainers are unlikely to agree with any of the above. Simple self-interest and tribal loyalties compel them to believe that all resources directed towards their sports or entertainment is justified, and the more the better.

The issues arising from the distraction and manipulation of the masses through sports is a peculiarly human phenomenon.

The human sex drive

Sex is about gene propagation, sometimes referred to as the principle of the 'selfish gene.'[31] Human males and females are unconsciously primed to maximise gene propagation and ensure the continuity of their own genetic line.

Maslow's hierarchy places sex in the needs category along with food and breathing; it includes sex solely from an individualistic perspective. Sex is placed with the other physiological needs to be satisfied before 'higher' levels of motivation come into play. Some critics of Maslow feel this placement of sex neglects the emotional, familial, and evolutionary implications of sex within the community. But sex is not a universal need, evidenced by the fact that children do not need sex, and adults can choose to go their entire life without it and yet can still fulfil higher needs. The same cannot be said for Maslow's other listed essential needs.

Sex is only relevant to those species that need to reproduce and then do so in a sexual manner. Sexual reproduction does not manifest some necessary universal principle or need. It is an incident of a particular evolutionary biological process that has occurred on the Earth and some other planets. It is an early casualty of any self-directed evolution and has no part in advanced alien civilisations.

The human male sex drive, when suppressed, emerges in alternative forms of competition or violence at both the individual and collective levels; for example, as road rage, in pub fights, as domestic violence, or through wars. The female sex drive rarely manifests in this manner; as an example, women are almost never the perpetrators of mass shootings.

Historically, women have been socially programmed to 'nest,' to attempt to find an emotionally and economically stable partner, and to protect their offspring. These mechanisms helped ensure not only the propagation of their genes but also the survival of offspring to maturity so that they too can propagate. These assertions are often criticised as gross generalisations, and reference exceptions that include the existence of people (including men) who have no desire for children. But again, the exception does not negate the general rule.

The drive to protect and propagate one's genetic heritage and offspring is a major motivator for humans of all genders and sexual orientations.[32] This need accounts for much of the actual sociological conduct on Earth. By way of example, it explains the peculiar parental behaviour known as 'Mengele's Mother Syndrome'[33]. This is a term coined to describe parents, particularly mothers, who are unable to accept that their child may have engaged in wrongdoing or might be an obnoxious bully. To this type of parent, the child is innocent and misunderstood. Again, obviously there are exceptions.

Genetic loyalty operates both downwards and upwards i.e. from parent to child and from adult child to parent. The programming to protect an individual's gene pool underpins a distressing spectacle in nursing homes. Some residents' physical, mental, psychological, and neurological health may have deteriorated to the point where they demonstrate little or no self-awareness or visible enjoyment of life. In the worst cases these people may just sit in a chair all day and stare into space, with their mouths open, drooling and moaning. Some families resist bringing an end to the misery and indignity endured by their suffering family members, including actively resisting euthanasia. Instead, these individuals are made to endure this misery to placate the misguided and entrenched genetic loyalty and associated guilt of the family. In turn, society must allocate the significant resources to support these individuals—resources that could have been diverted elsewhere, for more productive outcomes for more people.

Part of the motivation for resisting voluntary euthanasia is the implicit myth that all humans of a certain class are entitled to live as long as possible, with as little suffering as possible. In the human context, this is just not true. Moses did not descend from Mount Sinai with an eleventh commandment of 'Thou shalt live long and prosper.' The only humanoid species that subscribes to this idea is the fictional Vulcans of *Star Trek*. The accidental (and fictional) American philosopher, Forrest Gump, encapsulated the contrary argument on his T-shirt. It read, 'Shit happens and then you die.' People should be allowed to die with dignity at a time of their own choosing.

Human sexuality is a consequence of the form of reproduction adopted during the evolutionary process. It was not a necessary path through evolution. Asexual (or non-sexual) reproduction in humans could have arisen. Had evolution taken this path, then there would be no genders and no concept of sexuality to obsess over and motivate humans. The social customs and values that arise from the existence of sexuality would not exist. What then would human society look like?

Physical beauty as a motivator

Related to the issue of the sex drive is the concept of physical beauty. The idea of 'beauty' referred to here is not that of the artist or scientist, which would be the aesthetics arising from the marriage of form and function. It is the perceived, subjective physical beauty of individual human persons. Human physical beauty is perceived differently across cultural divides. This may explain in part the evolution of different physical traits. Humans with more desirable attributes are likely to be desired partners and thus the desirable traits (e.g., symmetrical face, big eyes, large chest, et cetera) are more likely to be reproduced in any offspring. Once a given physical characteristics is preferred in a society, people with these characteristics gain an evolutionary advantage over others. They mate more and produce more offspring.

What passes for physical beauty among humankind does not register with aliens. Aliens judging human-desirable physical features such as height, weight, skin colour, and hair colour and texture, may be akin to humans attempting to assess the physical attractiveness of one crocodile to another, by reference to characteristics of their skins or scales or the sharpness of their crocodile teeth.

By way of illustration, in the movie *Planet of the Apes*, a human interloper, played by Charlton Heston (1923–2008), formed a bond with a scientist who is a female ape. The latter felt that their bonding was adversely affected because the human Heston 'was so ugly.' Physical beauty is a sociological construct and will not survive the changes in a society.

Huge economic and social resources are directed towards the worship of physical beauty by humans. It underpins the cosmetic, fashion, film, music, advertising, and other industries. Beauty is used to sell just about everything, from cars to toothpaste and dishwashing detergent. It is used to body shame people and to establish tribes of beautiful 'influencers' who feature among the contemporary 'heroes' of society.

Many young people, but mostly females, take selfies—pictures of themselves—in public. Usually these are taken after primping and preening and before being airbrushed and filtered for posting to social media sites. Young males in the gym spend an inordinate amount of time gesturing and admiring themselves in the surrounding mirrors. This is reminiscent of the Greek mythological character, Narcissus, who fell in love with his own reflection, seen in a pool of still water.

Recently, a news article showed a female celebrity wearing a famous dress once worn by Marilyn Monroe (1926–1962) when she sang *Happy Birthday* to President John F. Kennedy (1917–1963) in May 1962. The article noted that the celebrity paid US$7 million to hire the iconic dress for one wearing. The dress was apparently damaged, and the celebrity reportedly had to pay to remedy that. This raises the question of whether narcissism should be subject to any moral constraints. Is it acceptable to say, 'the money is mine and I will decide how to spend it'? If an affirmative response is broadly implemented, where does such moral philosophy take humankind in the future and in terms of converging universal ethical norms? It is simply licensing the ongoing polarisation of humanity into 'haves' and 'have nots.' It is acceptable for many to starve whilst others grow obese, because the obese have the right to spend their money as they see fit and waste what they choose to waste.

What, ultimately, is the value to humanity of physical beauty? What does it add to the progression of any evolutionary advantage to the species? It is a mere temporary, socially conditioned expression of preference for one physical characteristic over another. It says nothing about the intellectual or moral character of the individual or the species. Unless it retains some evolutionary benefit, it will pass out of existence as the human form evolves.

Toxic masculinity and war

Human males are subconsciously, sociologically, and chemically compelled to strive to be the alpha or dominant male in the human herd. To this end, males seek power over other males and to own or possesses females. The latter is the origin of the institution of marriage. To do this, males utilise their available competitive power while seeking to minimise the impact of any disadvantage. That advantage might be physical, economic, or political. They will consciously or subconsciously promote social systems that minimise their disadvantages.

This simple point was made in the opening scenes of Stanley Kubrick's (1928–1999) film *2001: A Space Odyssey*, where two tribes of an ape-like species confronted each other over the possession of a water hole and the chance to mate with a particular female ape. One of the male apes picked up a large leg bone and realised he could use it as a weapon to beat the males of the other tribe. In the end, his tribe won the water hole and the female. This was not doubt a true depiction of the early days of human evolution. Since then the weapons have evolved, and the fights extend across the globe, but the underlying motivations of power and greed, and the use of savagery to acquire the desired items remain common.

Historically, monarchs, feudal aristocrats, and the warrior classes of human societies almost always developed from this motivation. It is the sublimated male sex drive that causes almost all conflict, whether great or small. It is historically that same psychological driver that underpins forms of autocratic behaviour including everything from military dictatorships, the capitalist economic system's pursuit of economic dominance, to the level of daily domestic

In Defense of the Alien

violence. This form of aggression is sometimes called 'toxic masculinity.'

An old but good episode of *Star Trek* made this point. The starship *Enterprise* was on a mission near Earth when it found a cryogenic cylinder floating in space, which contained a man from the 20th century. He had been a rich man who had sought to engineer his own immortality via this cryogenic technique, coupled with the expectation that he would be found and revived in a future time when his medical issue could be attended to. After he was revived, he spoke to the *Enterprise*'s Captain Kirk.

The man asked what business opportunities were available on the *Enterprise*, on Earth, or elsewhere. The captain did not understand the question and asked what kind of activities the man was referring to. The man said that he was looking for opportunities to generate wealth. The captain asked, 'Why, what is it that you want that you don't have?' The man replied that he wanted 'power,' to which the captain responded, 'What kind of power, for what?' The man said that power over others is the mark of a man, and attracts the best in life, including women. The captain responded that there was limitless food and accommodation available for free; the man could go to Starfleet Academy and train to be a captain and command a starship; there were as many women as men on the *Enterprise*, Earth and elsewhere, with whom he was free to strike up a relationship. The man responded that he would then have to compete with other men for both the captain's position and the women. He said something to the effect of, 'But I don't want to compete with other men; I want to eliminate my competitors.'

The captain walked away and spoke with the *Enterprise*'s psychologist. She explained to the captain that this was the mentality of men in the 20th century, and was considered to be a good and honourable motivation and conduct at the time.

This form of toxic masculinity is pervasive and has been present throughout recorded human history from the time of the prehistoric cave dwellers illustrated in Kubrick's opening scenes. It is why there are constant geopolitical threats of war, now including war in space. The sabre-rattling threat of war is one way that toxic masculinity expresses itself.

Rhetorically, one could ask: When has a war ever occurred based on one group of people expressing concern and a desire to help or share with another group? When has a war occurred because one group of people have sought genuine dialogue with another over some contentious point? In almost every case, war arises because men in a position of power develop their own private agendas, whether conscious or subconscious, that are based in toxic masculinity.

By way of example, the quote attributed to Julius Caesar (100 BCE–44 BCE) following his conquest of the British Isles goes: 'I came, I saw, I conquered.' He did not say: 'I came, I saw, I befriended' or anything similar. His purported statement simply assumes that anything new is to be subjugated or killed—much like the poet John Keats's (1725–1821) hapless nightingale sitting and singing in a tree. A human passing by simply assumed that the bird was there to be killed; why else would it be sitting in the tree?[34]

The famous English historian and linguist, Dr Samuel Johnson (1709–1784), once wrote:

> What are all the records of history but narratives of successive villainies, of treason and usurpation, massacres and wars. ["The Rambler' 1751]

Some wars are the outcome of expansionist land grabs.[35] Some wars are grabs for resources.[36] Some wars are for political or religious ideological purposes, including regime change.[37] Some are presented as wars of liberation, unification, or revolution.

The fact is that humans have been in a constant state of war of various kinds, across the planet, since the recorded history of humanity and earlier. The point is sufficiently illustrated by the 19th and 20th centuries, but is not limited to these centuries. The 19th century started with the Napoleonic Wars (1803–1815) and ended with the Boer War (1899–1902) between Great Britain and The Netherlands for control of South Africa. In between those two conflicts, there was not a year that passed where there was not a war occurring.

In the 20th century, apart from the two World Wars, the Korean War and the war in Vietnam, there were numerous wars of 'liberation' in Africa, Asia, and Europe, and genocidal conflicts that have kept the United Nations Peacekeeping Forces constantly deployed.

In the 21st century, so far there have been wars in Africa, Asia, and Europe, including genocides, and there is expansionist sabre-rattling from China and Russia—no doubt with more wars to come.

The reality is that men will always find a reason to go to war. Asking men why they chose war is akin to asking the cat why it chased the mouse. Aliens observing humans just see a continuous pattern of resorting to violence to resolve disputes. Aliens see humans in much the same way that humans observe ants and their constant fighting on the ant mound.

It is not just that human history is one of continuous war, as noted in Samuel Johnson. It goes further; many human males think that war and conflict is a good thing! Some examples in chronological order: the Catholic monk, Honoré Bonet (c.1340–c.1410) in 1390 once wrote:

> War is not an evil thing, but good and virtuous; for war, by its very nature, seeks nothing other than to set wrong right and turn dissention to peace, in accordance with the scripture. And if in war many evil things are done, they never come from the nature of war but from false usage.
> ["The Tree of Battles" c. 1382-87]

John Stuart Mills (1806-1873), a principal proponent of the philosophy of 'utilitarianism' (the greatest good for the greatest number) wrote that whilst war was ugly it was sometimes a necessity and in this sense was a virtue for those men who for fought for the good cause:

> "War is an ugly thing, but not the ugliest of things: the decayed and degraded state of moral and patriotic feeling which thinks that nothing is worth a war, is much worse. When a people are used as mere human instruments for firing cannon or thrusting bayonets, in the service and for the selfish purposes of a master, such war degrades a people. A war to protect other human beings against tyrannical injustice; a war to give victory to their own ideas of right and good, and which is their own war, carried on for an honest purpose by their free choice, — is often the means of their regeneration. A man who has nothing which he is willing to fight for, nothing which he cares more about than he does

about his personal safety, is a miserable creature who has no chance of being free, unless made and kept so by the exertions of better men than himself. As long as justice and injustice have not terminated their ever-renewing fight for ascendancy in the affairs of mankind, human beings must be willing, when need is, to do battle for the one against the other."
("Principles of Political Economy" 1848)

A United States president and one of the founding fathers of that country, Thomas Jefferson (1743–1826) wrote in 1787 in his 'Tree of Liberty' letter concerning the need for armed rebellion:

God forbid that we should ever be 20 years without a rebellion ... if they the people remain quiet under such misconceptions it is a lethargy, the forerunner of death to the public liberty ... The tree of liberty must be refreshed from time to time with the blood of patriots and tyrants. It is its natural manure.

Just before the First World War, the Prussian general and military historian, Friedrich von Bernhardi (1849–1930) wrote:

War is a biological necessity of the first importance, a regulative element in the life of humankind which cannot be dispensed with ... But it is not a biological law but a

In Defense of the Alien

moral obligation, and, as such, an indispensable factor in civilization.

['Germany and the Next War" 1911]

Von Bernhardi may have had in mind the argument of Thomas Malthus (1766–1834) who argued that war and disease were logical and natural responses to overpopulation. Malthus did not say that people decide to go to war for the purpose of reducing overpopulation, or that they deliberately cause pandemics for that purpose. Rather, the competition for the resources needed for survival creates the subconscious drivers of war, and the close physical proximity of overpopulation promotes the development and spread of disease.

Sir Ian Hamilton (1853–1947) was the commander of the Australian and New Zealand Army Corps (ANZAC) that suffered in the unmitigated disaster at Gallipoli in 1915 (on a date now celebrated as a proxy for Australia's national day). Gallipoli involved an attempted invasion of Turkey in an ill-conceived and poorly executed brain snap of the British aristocrat Winston Churchill (1874–1975), then First Lord of the Admiralty. Tens of thousands of men and boys were ordered ashore to charge up the hills and cliffs at Gallipoli, against entrenched machine-gun fire and shelling, whilst Hamilton remained safely distant. Hamilton wrote:

Albert de Alien

> There are poets and writers who see naught in war but carrion, filth, savagery, and horror ... They refuse war the credit of being the only exercise in devotion on the large scale existing in this world. The superb moral victor over death leaves them cold. Each one to his taste. To me this is no valley of death - it is a valley brim full of life at its highest power.
> [Gallipoli Diary Vols 1-11]]

Hamilton's attitude may have been common for men at that time, but its logical and moral offensiveness is hopefully now self-evident to many and was the subject of trenchant criticism by the First World War Poets and was always offensive to aliens. Some men may have since moved on from sharing Hamilton's view, but not all. The same blunt and sacrificial, 'crash or crash through' military attitude and consequential slaughter is still occurring elsewhere in the world at the present time.

The Gallipoli mentality bears analogy to the actions of the old men who constitute the mullahs of Islam, who urge young men and women (and even children) to strap on bombs and kill themselves and others, all in the cause of a just god, whilst the mullah remains safely distant. What would make these old mullahs more convincing would be an event in which they all came together in an arena with bombs strapped on themselves, and demonstrated their faith by systematically blowing themselves up, without causing any harm to others. It is doubtful thought that their faith extends this far.

The concept of a 'just war' goes back to St Augustine (c. 354–c.430), in his book *The City of God* written circa 413 CE. It sets out a range of criteria for when it is just to go to war and what the rules of engagement should be. Self-defence and the defence of the life of others are two such reasons. So is the defence of one's own sovereignty or that of another. It may therefore be just to go to war over nationalist land disputes, or to resist or assist attempted regime changes. History usually assesses the justness of a war from the perspective of the victor, who inevitably writes that history.

There is a tendency for humans to rationalise their participation in these wars as being fights to preserve their freedom. It is simply not true that a country, or tribes within a country, that participate in these wars, or are part of the United Nations Peacekeeping Forces, are necessarily doing so to protect their own freedom or the freedom of others. It may or may not be the case; it depends from whose perspective the matter is assessed.

The Japanese might not accept that WWII was a war of aggression on its part. Americans, British and Australians would almost certainly not accept that the Second Gulf War was a war of aggression motivated by regime change in Iraq, that could have been addressed in other ways. They would not accept that it was a naked display of power by President Bush following the 9/11 disaster, for the purpose of placating domestic anger, with UK Prime Minister Blair and Australian Prime Minister Howard dutifully tagging along.

Serbians are unlikely to agree that the United Nations Peacekeeping intervention in the Yugoslavian Wars (1991–1997) were interventions to preserve the freedom of Croatians, Bosnians etc. Russians might not accept that its invasion of Ukraine was just a naked land grab.

Some people have sought to oppose various wars. Bertrand Russell (1872–1970) was one of them; the UK Astronomer Royal, Arthur Eddington (1882–1944), and the world's most influential physicist, Albert Einstein (1879–1955) were others, vocal in their opposition to WWI. Their arguments and those of others fell on deaf ears, and they were pilloried as cowards and traitors. It was not until after the horror and pointlessness of WWI had played out that public opinion on the war changed. This point was poignantly made by the poets of WWI and in the opening scenes of the film *Chariots of Fire*.

History repeated itself with the Vietnam War. It is now almost universally recognised that the Vietnam War had been ill-conceived and poorly executed. Even Robert Strange McNamara (1916–2009), the US Secretary of Defence responsible for escalating the Vietnam War, made this

In Defense of the Alien

admission in an interview shortly before his death. People in the US and Australia who spoke out against the Vietnam War at the time were pilloried as cowards and traitors by the majority who marched to the beat of the drums of nationalism and war.

History has vindicated those who protested against the First World War and the Vietnam War. History has shown how shallow and easily manipulated was the great bulk of the population that supported these wars, and this is still the case. History levels the same accusation against other human populations that are responsible for other wars, including against the present Russian people who support President Putin's invasion and slaughter in Ukraine.

There is a lesson for humans to be learned from this history. Beware the power of tribal propaganda. Beware accepting as fact the narrative of governments, media, social media, or anyone else that has a possible self-interest in the issue at hand, or who has an apparent lack of a sufficient knowledge base to support their narrative. To 'beware' (or 'be aware') of something betokens a critical and sceptical approach, it is not license to simply summarily dismiss a narrative. The American philosopher Howard Zinn (1922–2010) once put it thus:

> I'm worried that students will take their obedient place in society and look to become successful cogs in the wheel - let

> the wheel spin them around as it wants without taking a look at what they're doing. I'm concerned that students not become passive acceptors of the official doctrine that's handed down to them from the White House, the media, textbooks, teachers and preachers.
> [https://www.inspirationalstories.com/quotes/howard-zinn-im-worried-that-students-will-take-their]

The members of the military that have participated in various wars often argue that they just obeyed orders and were unaware of the full context that is later revealed as history. Even at just the human level, the Nuremburg trials after WWII, and later international war crimes trials, have made it plain that merely obeying orders is not a justification for committing crimes against humanity, and this principle is consistent with the universal ethical norms referred to above.

History shows that in general the people involved in these wars were not interested at the time in knowing the facts or context; they were wholly consumed in the nationalistic propaganda and had turned off any moral compass they might have had. Others go to war because it is their opportunity for adventure, to prove themselves, or to advance their military careers. Others support the wars because it is a source of profit for them, through supplying materials and developing lethal technologies.

In Defense of the Alien

But once a soldier has adopted the military narrative, they then see their actions as part of their duty to their country. Within limits, it is not right to pillory military personnel at a personal level. The criticism should be levelled at the political and media leaders who instigated the nationalist manipulation in the first place. Within limits, and at the personal level, the actions of the individual soldier may be justified on this basis and can be heroic in their own way.

For example, consider Homer's classic epic poem, the *Iliad*; most readers would not condone the actions of Paris of Troy in abusing the hospitality of Menelaus, King of Sparta, by absconding with his wife, Helen. Nor would they see the defence of Paris by his brother, Hector of Troy, during the ensuing siege of Troy, as being justified. In the end, Hector was aiding and abetting the crime of his brother. Hector was just adopting a form of Mengele's Mother Syndrome. He should have ordered the return of Helen, and thereby avoided the war that led to the massacre of most Trojans. Nevertheless, few would deny the heroism and pathos of Hector's brave, and inevitably fatal, personal defence of his brother outside the walls of Troy, when he fights against the overwhelming power of the demigod, Achilles.

Even after a war has started, the ordinary soldier does not have to obey the commands of a dictator or tyrant, or in fact any command that is plainly offensive to the universal ethic or their own moral compass. They may choose to do so for their own self-preservation, for example by rhetorically asking themselves, 'Is it better that I kill 100 innocent people, than I be killed or otherwise harshly dealt with for disobeying an order?' How far could Putin have got in Ukraine if the Russian soldiers had just said 'no' to his orders? This is the problem the singer, Donovan (b.1946, age 76), wrote and sang about in his 1970s classic song, 'The Universal Soldier' and nothing seems to have changed since:

> He's five foot-two and he's six feet-four
> He fights with missiles and spears
> He's all of 31 and he's only 17
> Been a soldier for a thousand years
>
> He's a Catholic, a Hindu, an Atheist, a Jain
> A Buddhist, and a Baptist, and a Jew
> And he knows he shouldn't kill
> And he knows he always will
> Kill you for me, my friend, and me for you
>
> And he's fighting for Canada
> He's fighting for France
> He's fighting for the U.S.A
> And he's fighting for the Russians
> And he's fighting for Japan
> And he thinks we'll put an end to war this way
>
> And he's fighting for Democracy
> He's fighting for the Reds
> He says it's for the peace of all

He's the one who must decide
Who's to live and who's to die
And he never sees the writing on the wall

But without him
How would Hitler have condemned them at Labau?
Without him Caesar would have stood alone
He's the one who gives his body as a weapon of the war
And without him all this killing can't go on

He's the Universal Soldier and he really is to blame
His orders come from far away no more
They come from here and there and you and me
And brothers, can't you see?
This is not the way we put the end to war
[Wikipedia]

These criticisms are not new. This point was made poignantly by Jacques Offenbach (1819–1880) in the 19th century play, *La grande-duchesse de Gérolstein*. In this play Offenbach had the Duchess satirising the militaristic attitude of men of her society (the Franco-Prussians) so successfully that later French governments, including that of Napoleon, banned the play.

The American philosopher Harold Zinn made the same point, in the following terms:

> Civil disobedience is not our problem. Our problem is civil obedience. Our problem is that people all over the world have obeyed the dictates of leaders ... and millions have been killed because of this obedience ... Our problem is that people are obedient all over the world in the face of poverty and starvation and stupidity, and war, and cruelty. Our problem is that people are obedient while the jails are full of petty thieves ... (and) the grand thieves are running the country. That's our problem. [...] Historically, the most terrible things—war, genocide, and slavery—have resulted not from disobedience, but from obedience. [...] Protest beyond the law is not a departure from democracy; it is absolutely essential to it.
> [https://randythym.com/2015/10/01/civil-disobedience-is-not-the-problem-the-problem-is-civil-obedience/]

Human societies spend enormous sums on the instruments of war; sums that could be diverted to other, more productive uses. Australia recently signed contracts for submarines and military aeroplanes costing something near AUS$100 billion, and then abandoned the submarine contracts with the French government and paid massive damages for ending the contracts. No money was spent on building hospital ships that could sail to disaster zones anywhere in the world and provide useful humanitarian assistance. Even if Australia needed to take defence so seriously as to need these military vehicles, there are other ways of doing this rather than just spending huge sums on weapons of war.

For example, Australia could couple defence spending with hosting an international space agency. There are many countries that are not individually wealthy enough to participate in the space industry, although collectively they could do so. Other established space agencies may also be interested. Australia has the land and facilities in central Australia, and an almost endless capacity to generate solar and hydrogen power to host communities directed to this end. Cooperation and collaboration are good for world peace. The existence of a major international space agency located in central and northern Australia would collaterally present a deterrent to any invasion.

Toxic masculinity and political and economic manipulation

The effects of toxic masculinity referred to above are not limited to causing wars. They also extend to commercial and political activities. It is why some businessmen—and far fewer businesswomen—take the view that there are no ethics in business; the only goal is the pursuit of profit and power, and to avoid detection of any legal transgressions. They are prepared to damage the environment, incentivise slave labour, avoid taxation, bribe public officials, destroy the livelihood of other people, and if economically convenient, encourage and profit from war or disasters. It is a repulsive spectacle to alien observers.

The famous industrialist Henry Ford (1863–1947) once wrote:

> There is something sacred about big business. Anything which is economically right is morally right.
> ["Moving Forward" 1930]

It is why it is not considered to be wrong or perverse for a small number of humans (mostly males) to grow physically and morally obese in their wealth, whilst other humans starve. It is not that such wealthy men can live in their 10 houses at the same time or eat the limitless food available to them or have sex with limitless women each day. It is the subconscious motivation that these things are symbols of power that betoken their virility. To surrender any of these assets would lessen them as a male, and further lessen their competitive advantage over other males.

It was not until the beginning of the 20th century that in most democracies, women were accorded full status in law as people and were allowed to vote. In other countries female suffrage was much later, and for some it still has not occurred. The arguments and demands for equal rights for women were made as long ago as 1791 in the *Declaration of the Rights of Woman and of the Female Citizen* by French playwright and activist Olympe de Gouges (1748–1793), in response to the 1789 *Declaration of the Rights of Man and of the Citizen* arising from the French Revolution. De Gouges hoped to expose the failures of the French Revolution in the recognition of gender equality, but her work failed to create any lasting impact.

In most nascent democracies, it was only in the 19th century that all men were allowed to vote, not just those from the wealthy classes (i.e., the aristocracy, landowners, clerics, and the new rich of the industrial revolution). People of certain religious faiths were still excluded from public office and from voting. Slavery in various forms still existed.

In present autocratic and theocratic societies, including tribal societies, men from dynastic lineage, ethnicity, wealth, and religious castes still control the societies and consciously or subconsciously jealously guard their political power.

Part 5: Tribalism as Both a Motivator and Consequence of Evolutionary Behaviour

Tribes and how they work

It is only at the third level of the Maslovian hierarchy that the need for social relationships arises. These social needs are not limited to humans, either. There are many animals that also form bonds of this description for life; whales, dolphins, and the family pet, for example. This is one reason why it is wrong to consider animals to be mere property and things to be owned, used, and killed by humans.

The need for communal or collective bonds and cooperation is also true of aliens. The difference is that for humans, tribes are actually a source of division or conflict within human society, rather than a source of cohesion. Tribalism is used as a strategic weapon by power-hungry men.

In Defense of the Alien

The human need for social relationships usually manifests in the need to belong to a group of some kind. This need has both positive and negative consequences. It is positive in promoting stable societies that can discourage some of the consequences of toxic masculinity, because violence and wars cause rupture of relationships. It is negative in that it exposes humans to psychological manipulation around their need for a sense of belonging and acceptance in a group.

Groups within a larger society go by different names, often referred to as 'tribes' or occasionally 'cults'. The tribes form around a core set of accepted beliefs, which permits their members to differentiate themselves from those not in the tribe or cult. They have a 'we and them' binary logic. People who are not in the tribe are against the tribe, in the sense that they are enemies of the tribe, or at least a danger to the purity of the tribe.

Jonathan Swift (1667–1745) poignantly made this observation in *Gulliver's Travels*. In this book, there was a dispute between two tribes over which end of a boiled egg should be cut for eating. They were prepared to go to war over this issue.

Tribes and cults can form around religious doctrines, perceived race/ethnicity, gender, wealth, country of origin, political affiliations, social agendas, and just about any other idea about which there can be conflicting opinions between people. People can also be a member of more than one tribe at a time. For example, the same person could be a 'white Christian nationalist' where there are three overlapping tribes. The same person may also be a 'misogynistic homophobic capitalist' and so be a member of six overlapping tribes at the same time. The past US President Donald Trump would be such an example.

Often the cost of admission to these groups is loyalty to a set of doctrines, or to the authority of a particular person (usually a male who is seeking dominance). People within tribes either consciously or subconsciously interpret external social reality in terms of the values system of the cult or tribe, hence why it is extremely difficult to have a meaningful conversation with them.

In order to gain acceptance within the tribe or cult and to remain therein, people will often convince themselves of the objective fairness and righteousness of the tribe's core doctrines. This is needed in order to avoid emotional upheaval within themselves stemming from cognitive dissonance—the internal discomfort brought on by having two conflicting thoughts or ideas—as well as to avoid expulsion from the tribe or cult. Cultural osmosis is the indirect process by which a person within a particular tribe or cult will absorb and come to believe its doctrines and agendas without necessarily having direct experience of the subject. Subconsciously, it is easier to swim with the tide than against it, and it is the price for membership within the group.

Tribal loyalties can be so strong that the attitude of a tribe member to any particular issues can be predicted without having to discuss the issue with that member, and without the need to bother about the facts. The equation is simple; if the proposition in question is consistent with the tribe's values and interests, then it is plainly a true proposition. Conversely, if it goes against the tribe's values and interests, it is plainly false.

It can and regularly does reduce to the ridiculous. Take a recent example, where the French president accused the former Australian prime minister of lying to him. In the Australian media and population, the split was simple. Members of the former prime minister's political tribe took the view that it was the French president who was lying. Political opponents of the former prime minister readily accepted that the prime minister had lied, and regularly did so. Almost no one was aware of the actual facts and content of the communications in question. The contest was resolvable on simple tribal grounds alone. This same phenomenon applies to just about any pronouncement made by political leaders.

The same simple equation applies to the question of whether former President Trump colluded with far-right elements in an attack on the Capitol Building in the US capital, in what amounted to an attempted coup. Trump opponents say it is a true proposition. Trump supporters say that it is a false proposition and that there is no need for any factual investigations on the issue, and no need to hear the evidence each way.

Tribes built around conspiracy

Tribes that form around conspiracy theories are usually based around knowledge ('gnosis' in Greek) in one form or another that is usually kept secret. The idea of 'gnosis' derives in part from the early Christian church schism called 'Gnosticism,' also derived from the Greek word. Gnostics believed that they were in possession of secret and special knowledge about the Gospels, Jesus, and God that other church members were not.[38]

Gnostic sects still exist, mostly in the form of closed cults or closed religious communities who indoctrinate their members with the belief that they hold the truth, while the rest of the world lives in a state of deception. A key characteristic of these communities is to prohibit their members from exposure to contrary points of view, and from engaging in their own thinking on relevant issues.

Independent thinking is taught to be an expression of arrogance on the part of the sect member, which is a sin.

This was something highlighted in Ray Bradbury's (1920–2012) classic novel *Fahrenheit 451*, set in a future where all books or other forms of recorded expression or learning were burned, along with the house in which they were found.

Present-day conspiracy theorists do the same thing. They have their own trusted sources of information whose pronouncements are taken as gospel or ultimate truth. There is to be no debate over the motivations of these trusted sources, the logic they employ, or their relevant expertise on a topic.

In the Protestant branches of Christianity, this sometimes comes down to the estrangement between different sects or churches over the interpretation of the same words of the New Testament. Each will vehemently assert that they have a personal relationship with God, who has personally assured them of the correctness of their interpretation of the Biblical text, and that those who propose the contrary are misguided, or even demonic. Recall the example of this given earlier, about the fall of Constantinople.

The dictator Adolf Hitler (1889–1945) once wrote:

> The great mass of people … will more easily fall victim to a big lie than to a small one.
> ["Mein Kampf"]

The dissent between groups that hold strong and contradictory views has now become a common source of violent civil dissent in democratic societies and even in some autocratic ones. In the US this trend is approaching the point where the US may soon become ungovernable. There is a superficial parallel between the reasons for the decline and fall of the Roman Empire and the unrest presently happening in the US. The main reason Rome fell was because of the loss of any unifying civic duty or authority.[39] Individual strong men could not resist attempts to obtain or hold onto power by coups and civil wars. The only method for the transference of power was the sword. Toxic masculinity was running rampant.

Internet sites that form the authoritative basis for some conspiracies are plainly commercial manipulations. They place advertisements or seek to sell their own materials or remedies of some kind. Some are run by people who have their own psychological issues, which cause them to seek attention or affirmation—ultimately the psychological drivers for the 'gnosis' principle. But the conundrum remains as to why otherwise-intelligent people would be so easily manipulated. The answer may simply lay in the psychological vulnerability of nearly all humans, and the success of these conspiracy theorists in tapping into that vulnerability.

In Australia recently thousands of people protested the requirement to be vaccinated against COVID-19. The protestors complained that their consent should be required, and that they did not give that consent. They argued that they, the protestors, were young and fit and therefore not much at risk from COVID. They rhetorically asked why they should suffer the inconvenience of being vaccinated. Yet the answer is simple, it is in the interests of the community as a whole for them to be vaccinated, including for the benefit of the older and the vulnerable. It is ultimately in the self-interest of the young and fit because it suppresses, though does not entirely eliminate, the spread of the virus.

In fact, at that time in Australia unvaccinated people, per capita, constituted by far the greater percentage of people admitted to emergency hospitals, where they apparently had no reservations about using the hospital's resources and endangering hospital staff.

The above rhetorical question by those opposing vaccinations has about the same logical and ethical persuasiveness as the rhetorical question once posed by Australia's then-richest man, Kerry Packer, when questioned on his tax evasion by the Australian Senate. He said:

> Why should I pay any more tax than I have to, given how the government spends it?

Shortly after this he had a heart attack, and called on the publicly funded ambulance service to convey him to a publicly funded hospital on publicly funded roads, for treatment by publicly funded medical staff, for life-saving emergency surgery. He did not decline these services. The ambulance in question fortunately was equipped with its own heart defibrillation machine, which was instrumental in saving his life at that time.

His subsequent response was to publicly announce that he would sponsor the purchase of heart defibrillation machines in all New South Wales (NSW) ambulances. He only deemed it fit to disgorge this small part of this vast wealth because he found himself in need of these public services. Even then, he felt the need to make this announcement to the public, as if he thought that this constituted some form of public atonement.

Religious tribes

Most if not all human societies, through the course of history, have developed a religion of some kind. Religion in general may be an attempt to grapple with the big existential questions of life, including personal and social ethics. But human religions suffer the handicap of being a human construct and are thus infected with human power politics.

For most religious people, their faith is something they are born or socialised into. It is rarely a position that people arrived at by a serious study of the content of the doctrines of a religion relative to that of other religions, or no religion at all.[40] For others, religion is only a social custom that is an indicator of belonging to a particular community.

The role and content of religions vary over time and between societies. It may be that the evolution of religions reflects the same kind of intellectual journey as does that of the physical sciences and moral philosophy. At any time in that intellectual journey, different societies have a partial understanding of the universal truths that are embodied in written or oral forms, consistent with their state of evolution at that time. This could be in the form of oral myths, sign languages, customs, ballads, songs, or written or symbolic language including things like mathematics.

There is often not much difference between myths, rituals, and religions. They may each embody something of the journey of humankind and its struggle to understand the universe. The difference is mostly one of form and emphasis, and one can merge into the other. Religions come in different forms, some overt and some not. Not all have a deity or belief in an afterlife. But they generally share the characteristic of having a set of firmly held common beliefs or doctrines.

Religions will generally have a 'guru' (spiritual teacher) who is the font of all wisdom and serves as intermediary between the people and their god(s). In the end there is not much difference between worshipping a purported disembodied spirit and worshipping an embodied person. A religion can even take the form of a cult of the state.

The doctrines or articles of faith of a religion mark the boundary of their tribe. For example, some Christian tribes assert that all human life is sacred; hence, they oppose abortions or euthanasia. Yet these same tribes have historically and in the present shown their willingness to engage in all manner of violence and atrocities in the name of their god. They have supported wars, just or otherwise, and opposed taxing of the wealthy to save the poor. They have condoned the abuse of innocent children.

By way of example, in 1981 the influential pastor and leader of the Moral Majority in the US, Rev Jerry Falwell (1933–2007), wrote:

> The free enterprise system is clearly outlined in the Book of Proverbs in the Bible. Jesus Christ made it clear that the work ethic was part of His plan for man. Ownership of property is biblical. Competition in business is biblical. Ambitious and successful business management is clearly outlined as part of Gods plan for his people. The bible belt is Israel's safety belt. ["Ten Ways the Free Enterprise Economic Model Aligns with Scripture"][41]

Former UK Prime Minister expressed a similar view.

This kind of theology originally springs from the sociology of Max Weber (1864–1920) and his argument that capitalism arose from a religious base in the form of the Protestant work ethic. The argument runs that each person had his/her allotted role or vocation in life and it was God's will that they labour earnestly in that vocation, be thrifty, and respect their superiors. The accumulation of wealth was a sign of God's approval of the person.

This theology has morphed into 'the prosperity gospel,' under which God shows its approval of an appointed leader by conferring on them opportunities for great material wealth, even if the opportunity comes in the form of the right to exploit their followers.

In the 1960s millions of Americans believed that God spoke through a 4-year-old child evangelist, Marjoe Gortner (b.1944, age 78). His parents conducted evangelistic crusades using him as a focus, that raised millions of dollars for themselves. As a young man, in his 1971 Oscar-winning documentary Gortner exposed the range of verbal and emotional tricks employed by himself and his parents to induce people to part with their money. The recent 2022 BBC World Service interview with Gortner presents an even bleaker picture of the ease with which religious people can be manipulated.

In 1987 the prominent US evangelist Oral Roberts (1918–2009) famously told his followers that unless they sent him US$8 million by the end of March that year, God would call him home—i.e., he would die. The followers duly complied. Many of TV evangelists are very wealthy people, and unashamedly so. There have been public scandals of evangelist couples making millions from gullible adherents whilst they live anything but an ethical life, and some of whom ultimately finished up in gaol. It is hard to see how the penny did not drop for the gullible people that were cheated.

Religiously motivated violence is not limited to Christianity. In Kabul in May 2021 a dispute between the Taliban and the Islamic State jihadist group led to the bombing of a girls' school, killing dozens of 11–12-year-old girls. It was an action authorised by old men behind the scenes in the pursuit of their own power and bigotries, presented to the public as actions on behalf of a just god. There are many such examples occurring almost daily.

One facet of human religions that is puzzling is the 'grovelling' approach whereby humans think that it is a duty imposed by their God that they must worship their God and do so in a grovelling manner. An advanced and ethical alien life form would not want this behaviour from humans or any less-developed life form. Again, it is just anthropomorphising a form of power desired by human, namely obedience and obsequious behaviour from other humans.

As a human, imagine walking out of your house to be met with a row of lower life forms such as birds and cats, all repetitively chanting in unison 'how great thou art.' It is akin to making your children into domestic slaves to stroke your own ego. A 'god' that wants this kind of obsequious praise is really just a reflection of the human ego and is not a god worthy of the title. The philosopher Galen Strawson (b.1952, age 70) by similar logic once said, 'it is an insult to god to believe in god'.

There is an obvious and historical problem with the concept of a prayer that petitions God to do a particular action, for example to break a drought or to end a war. Prior to the prayer either God thought that it was the right thing to do to break the drought or end the war, or God did not. In the latter case, it said that it was the prayer of the faithful that induced God to change its mind and end the drought or the war. The questions are: why would God change its mind simply because a faithful person asked? What about faithful people asking for mutually inconsistent outcomes such as both sides in a war praying for success? What does God do then?

During Easter 2022 the Russian president attended a service at the Russian Orthodox Church while the Ukrainian president attended a service at the Ukrainian Orthodox Church. Both prayed for divine support in their war against each other. The two churches were the same church prior to the onset of the Ukraine War, and so presumably both presidents were praying to the same God.

By what means could a person of faith ever determine whether their thoughts and prayers, and their perception of the responses to those prayers, were real, imagined or just a matter of their own confirmational bias? People who suffer schizophrenia have thoughts, hear voices, see visions, and are convinced of the reality of them until the episode is over, at which time they return to conscious reality. Most people have vivid dreams as well.

One possibility arises if the god in question is in fact an advanced alien capable of reading human thought and implanting thoughts in response. This alien would then be finite and capable of empathy and of a change of mind, but also may not have the ability to create or produce the outcome that is prayed for.

Religious tribalism presents a particular problem for liberal democracies in the world. It is all very well for people with autocratic political or religious agendas to plead for tolerance in a liberal democracy, and the right to proselytise. But once they become the dominant social group with control over that society, they become intolerant of dissenters and those that would seek a return to a liberal democracy or even simply a reciprocal tolerance.

This principle is most obvious in Islamic societies, where Muslim people deny the same tolerance to other religions that they sought for themselves when they were in a minority. Once in control, they discard the church/state divide and impose theocratic laws that are repressive to people of other faiths or of no faith. The civil war in Algeria (1991–2002) is such an example, where military and other non-Islamic interests declined to transfer power to an elected Islamic government for this reason.

But the problem is not limited to religious groups, as was shown by the shameful exhibition of Donald Trump refusing to surrender power and orchestrating the attack on the US Capitol Building, in a last-ditch attempt to hold onto power. If Trump could have held onto power and had himself appointed 'President for Life,' he would have done so.

Wealth-based tribes

Most people in the wealth-based tribe have convinced themselves that they have no duty to share their wealth with others. They seriously believe that they deserve their wealth because they have worked hard for it. The fact is that wealth in developed countries is in large part intergenerational; it is inherited, not earned. Nevertheless, these people believe that the poverty experienced by others is their own fault.

The idea is that someone in a developed country can make millions (or billions) of dollars by simply investing in non-productive speculations on the stock exchange e.g. futures or crypto currencies, and in so doing, is working harder than a man working 18 hours a day in the Mumbai slums to feed his family.

The poverty of a child born in a developing country is not the child's fault; nor is their lack of opportunity for health, education, housing, and employment. No one on the Earth is in a position to determine for themselves that they deserve any particular level of wealth or privilege over other people. One person's 'deserving' is another person's 'thieving.'

Unfortunately, the media dishes up a diet of this kind of self-serving nonsense on a regular basis. In particular, advertisers will tell people that 'you deserve' whatever is being advertised. Conservative politicians encourage all people to believe that such money as they have is 'your hard-earned money,' which they should resist parting with, in the form of taxes. It is obvious that this mantra is to the advantage of the people with the most money, who would otherwise have to pay the most tax, which would then be recycled as part of the 'social wage' for the benefit of people with lower income. Yet the propaganda is so successful that even people with lower income oppose any tax increases.

One common defence by the 'wealth tribe' is to assert that capitalism has proven to be the most efficient model at producing wealth for all, once the trickledown effect is taken into account. They point to the absence of any better example in recent human history, and in particular to the communist states of the 20th century. But these communist states were, and remain, simply military dictatorships imposed on the people from above.

Rarely do supporters of capitalism look at the boom-bust cycles that it creates, or other forms of market failure. They experience more agitation over the daily movements in share prices on the stock exchange than they do over the fact that more than half the world continues to live in poverty and disease, or with mass slaughters occurring in various places in the world, or melting ice caps. It is irrelevant to them that the weapons of the slaughters are provided by businessmen from developed countries, or that there is surplus food and medicine going to waste in the first world, or that current and historical industrial activity in developed regions is the main cause of the melting ice caps. The reality is that capitalism works well for those that profit from it but is destructive to others.

Albert de Alien

In Australia the recent housing market is an example of market failure. In the span of just a few years house prices spiralled upwards such that they more than doubled, including at the lower end of the market, while wages remained stagnant. First-home buyers, many young people, and people on lower incomes were locked out of the market. People with money to spare outbid these people at auctions and made large capital gains, which they took to be their right, and they saw no need for any government intervention or regulation.

The spiralling prices were for the same sets of bricks and mortar, which is pure inflation. There was no regulation in the market to prevent this, and the negative-gearing taxation actively promoted it. This movement dragged up rents for non-home owners. The absence of any market regulation caused the obvious and predictable outcome, as predictable as letting vampires loose in a blood bank.

In due course the housing market will probably fall dramatically, and then the banks and those that were all in favour of a free market approach at the time of their purchases, will be all in favour of the socialisation of their losses and expect the government to intervene to protect their investment. During all this time, the renters will continue to pay increasing rents. It is a wonderful system for those who profit from it, but only for as long as they profit from it.

In Defense of the Alien

There is no recorded European history of a social order that is not ultimately based on the survival of the strongest, and toxic masculinity, in one form or another. There were some limited attempts by Europeans to set up purported Utopias in the 18th and 19th centuries, which failed. There may have been some Indigenous communities in other places in the world that were not structured around survival of the strongest and toxic masculinity, but if they existed, then their social structure has not survived the effects of contact with Europeans and European religion.

This does not mean that a society based on a different social contract to that of capitalism, cannot or would not work better. This was the point of the *Star Trek* example given earlier. There are individuals who are imbued by non-capitalist goals, which shows that humans are capable of it. But it takes time and the will to evolve the concepts and consensus to achieve such a community.

If there has never been a recent and observed example of such a society, then it is irrational to assert the merits of capitalism based on the absence of any opposing example. This is essentially the logic that was used to support the continuation of slavery, i.e., it had shown its economic value, had always existed in one form or another, and there were no contrary examples. There have never been attempts to implement all manner of changes to society, but it does not logically follow that none of these changes are possible or desirable for the longer-term future of humanity.

It is a common experience bordering on a cliché, that many people in their youth have little wealth but lots of idealism. As their wealth accumulates, they progressively re-evaluate their idealism and assert that it has matured. There is little doubt that most people see more of the complexities of life as they age, but this of itself does not account for the consistency in which the maturation process correlates with economic self-interest.

On one view this maturity is the process whereby cultural osmosis has done its work, with the person progressively re-evaluating their belief system to facilitate their wealth acquisition, which is then something they will fiercely protect. This is not only true of individuals but also of institutions. How many governments, businesses and churches started out with philosophical or spiritual idealism, but over time evolved into businesses with assets to protect?

The tragedy of the ship *Titanic* provides a further metaphorical and real example of wealth-based tribalism. The wealthy people lived on the upper decks while the poor people lived on the lower decks. The ship hit an iceberg. There were not enough lifeboats for everyone, so the captain had those on the lower deck locked down until the wealthy passengers on the upper decks were evacuated. Once all the lifeboats had been launched, the lower deck people were allowed up. They had little choice but to jump into the freezing water. There were lifeboats that were only half full and were rowing away. Some people in those boats suggested turning around and picking some of the people in the water. However, with one exception, the lifeboats with their wealthy occupants decided not to risk being swamped with people from the water and rowed away to save their own lives, leaving the poorer people in the water to drown.

The one exception was a boat in which a woman attacked the masculinity of the men cowering in the lifeboat for their decision to leave women and children in the water to drown, just to save their own lives.[42] That boat turned around and picked up survivors, and contrary to the general belief, it was not swamped by doing so.

The future of humanity cannot involve the continuation of this kind of thinking and social structure. It is as primitive as the ant mound and, unless it is abandoned, humans in due course will meet the same fate as the ants and their mound. The human and other resources, real and potential, that are squandered in this highly greed-based system would be far better allocated to solving collective problems on the Earth, evolving helpful technologies, and exploring the solar system. Over the centuries, how many real or potential Ramanujans (a noted mathematician) or Einsteins have gone undiscovered or have been lost to the Earth? Take for example the case of the theoretical physicist Karl Schwarzschild (1873–1916), who was first to produce a solution to the field equations of Einstein's General Relativity. He did this whilst in the German trenches during WWI, where he was eventually killed—what a waste.

Nationalist tribes

The nationalist tribes are particularly dangerous to human evolution. They have proved easy to manipulate. History has shown how simple it is to indoctrinate the people of one nation to believe in their right and superiority over others, and to develop a sense of grievance, real or imagined, with other groups. All it takes is a little further manipulation, misinformation, and a little emotional jingoism by beating the drums of war, to achieve the desired militaristic end. Nationalist tribes often substantially overlap with racist, religious, or other tribes.

In Australia there has been a refugee issue for decades arising from the mess that Australia contributed to in the Middle East. A small number of these people have sought refuge in Australia by travelling in small and often unsafe boats, hence the term 'the boat people.' Australian politicians of all persuasions, and the media, have demonised these people in various ways and represented them as a threat to the country.

The governments and media have grossly misrepresented the total number of boat people, and the small number compared to the total migrant and refugee intake each year. They have developed various 'trigger words' contrived to elicit a calculated response in the population, including such terms as 'secure border,' 'sovereign borders', 'national sovereignty,' 'queue jumpers,' 'people traffickers,' 'operation sovereign border,' etc.

Until recently, they had for a decade or more adopted a policy of keeping these boat people in indefinite detention on offshore islands, or in detention camps on the Australian mainland. Some have been in these camps for many years now. Successive governments banned access to the camps to media, social workers, doctors, lawyers, advocates and anyone else that might report the truth. Telling the truth about the condition of these detainees was made a crime, but the bulk of the Australian population saw no problem with this. They preferred not to know, echoing the common response of a majority of German people during WWII to the existence of concentration camps. The principle is the same as is the moral status of the responsible Australian politicians.

Even if there were truth or merit in Australia's refugee policy, it does not justify the inhumane manner in which Australia has treated these people, including children.

This is simply the *Titanic* scenario playing out again. Australians are like the wealthy people in the lifeboats, who have the capacity to safely accommodate the boat people. But they choose not to, based on tribalism relating to nationality, religion, and race, with a good measure of greed added, in the form of an unwillingness to share. Politicians are attuned to the electoral value of this tribal bigotry, and develop various rationales to justify it, including the fiction that the boat people are 'queue jumping,' as well as the ubiquitous but surreal defence of the 'flood gates' being opened.

When viewed from a global perspective, this conduct is just more of the primitive ant-mound behaviour referred to earlier, and that portion of the Australian population that supports these policies is behaving like the automaton-resembling ants.

Part 6: Economic and Social Inertia of the Dominant Social Group

Once a dominant social tribe has secured control in a society, it is resistant to change unless that change is clearly to the advantage of that tribe. In liberal democracies this inertia can manifest in the political process. In autocracies it is maintained by the subjugation of the population by force or by the perversion of the rule of law.

By way of example, the policies of successive governments in Australia have been limited by the length of the electoral cycle, because this period is what is relevant to a politician's or government's re-election. The political debates have centred on how the existing economic pie is to be carved up between competing interest groups, rather than on growing the economic pie as a whole. Most of the existing wealth in Australia derives from what is dug out of the ground or grown in the soil, with a small amount of capital inflow.

Due to the relatively small population of Australia, this wealth results in a high per capita income compared to other countries in the world.

Notwithstanding this, the Australian population is told and believes that its affluence is 'hard-earned'. This has become an article of faith for the 'national tribe,' without having asked the question: 'hard-earned relative to what or to who?'; relative to the labourers of the developing world, or even to the working poor of the United States?

The continuing use and dependence on petrol and cars is a case in point. Petrol is not a renewable energy source; it is polluting; and it provides a source of geopolitical power on Earth. Technologically, it could have been dispensed with decades ago.

In the early 1970s *Dr Who* warned humans that their dependence on the 'black sludge' dug out of the ground was unwise, and that humans would come to regret it. At that time sufficient knowledge existed to have moved to hydrogen power. Every high school science student at that time knew that electrolysis of water produced hydrogen and oxygen, which could then be recombined to release heat. The *Apollo* and other spacecraft at the time used hydrogen- and oxygen-based fuels.

The decision to stay with petrol as the preferred fuel for cars was an economic choice, but not a technical necessity. It is often the case that technological solutions will follow the economic decision to pursue a solution, but unfortunately the converse is also the case, i.e., adverse economic decisions can stifle technological innovation.

In the last three decades of the 20th century humanity did not respond to the warnings by *Dr Who* and others about the dangers to the environment. Environmentalists were written off as 'extremist' using different pejorative labels. Again, the messenger was shot. 'Responsible' businesspeople ploughed on with their profit-producing projects, and the climate slowly started to change.

The history of climate change has shown that these 'extremists' were right and that the responsible businesspeople were wrong and anything but 'responsible.' They were in fact simply motivated by the immediate prospect of profit and power, and were not concerned with and did not want to know about any adverse environmental legacy they might leave behind. The responsible businesspeople of that time enjoyed their profits and are now gone, leaving a new generation to shoulder the burden they created.

The same is happening again today. Australian politics still struggles with the climate-change deniers, who are still only concerned with the short-term financial and electoral profits from coal, oil, gas, and other activities that objective science show to be dangerous to the climate. The fact that some people will lose their jobs in the fossil fuel industry is not a reason for maintaining a harmful industry. It is a reason for planned and regulated relocations of the emerging new industries.

Part 7: Maslow's Higher-Level Needs: Self-Esteem and Self-Actualisation

The fourth level on the Maslovian hierarchy is 'self-esteem.' This may be the first level at which human are differentiated from other animals, but it is a level that generally receives little prominence in human society. Self-esteem can be as simple as a sense of being politically or physical powerful, or financially prosperous, or can be the satisfaction of believing you are being true to your beliefs, whatever they may be and whatever the truth is.

For most humans, or more particularly males, there is a simple equation: self-esteem equals power, which equals economic wealth and the capacity to control others. For them, there is no point in having power or wealth unless it can be used. To these people there is an essential contradiction between possessing power and not using it to your own advantage.

Some people and some groups of people generally[43] are motivated to help others and to make the world a better place for all. But as with any general rule, there are exceptions, and exceptions within the exceptions. Not all people in these groups share this motivation. There are some whose motivation is also that of monetary profit, political posturing, or personal power of some kind, including so-called 'virtue signalling'.

Not all activities that provide self-esteem are productive or meaningful activities. Antoine de Saint-Exupery's (1900–1944) *The Little Prince* contains an entertaining story of how humans can strive for self-esteem or self fulfillment in the activities they choose, but the chosen activity can be an unproductive and meaningless one. He gives a range of examples including the example of a man who would be a king, who is content with the title, and theoretical power but has no subjects to rule over.

The real issue is that these people generally do not have the political or economic power in a society to be able to affect the evolution of that society's values or ethics.

The highest level of Maslovian needs is that concerning 'self-actualisation,' which is about the ethical, intellectual, spiritual, and creative pursuits of individuals within the social context. The number of humans engaged in these

In Defense of the Alien

pursuits is too small to be a significant factor in human development.

This need assumes that the lower-order needs have been largely satisfied. For most humans, this is not so, hence why this higher order need is essentially irrelevant to them. However, for advanced aliens the lower-order needs were long ago met via the choices in made in self-directed evolution, leaving available the pursuit to self-actualisation. Hopefully, the future of human evolution will take this same course.

Part 8: The Role of Language, The Need for a Common Human Language, and The Need for a Language for Communicating with Aliens

The role of language

Humans think, speak, write, and communicate by reference to one or more languages, but this is not true of all species. Most humans do not realise how complex their languages are, how central they are to every human society, and to the development of human thought and learning generally. Languages contain and conceal a cacophony of communal and personal values. Just about every non-trivial expression contains explicit and implicit assumptions. The speaker may or may not be conscious of those assumptions in their choice of language; as well, the speaker has no control over the

In Defense of the Alien

assumptions or implications that the hearer takes from, or applies to, the interpretation of the received statement.

Words and sentences are just symbols used to evoke or provide a meaning to the reader or listener. Neither the written nor the spoken word is self-interpreting, and so is not capable of delivering unambiguous communication. Ambiguity can also arise from the use of different linguistic devices, such as the use of metaphors, analogies, satire, sarcasm, jargon, irony, and implication. The more complex the content of the communication, the more ambiguous is its reception.

A statement may be grammatically correct, but does not necessarily contain any logical content. For example, poetry can evoke different feelings in different people, but still have no factual content. Some statements can be constructed of strings of elaborate language, and still convey no meaning. One human recently drew my attention to the growth of a new trend in clothes fashion called the 'neo-retro-definitive-ambiguous look' that I found perplexing.

It is a fiction to assume that every grammatically correct statement conveys meaningful content. The misuse of language in this manner is made into an art form by politicians. They are the tools of the trade for propagandists, politicians, advertisers, and conspiracy theorists.

Albert de Alien

There are particular oratorial techniques used by some people to exploit some aspects of language. One such technique is to speak rapidly at an elevated volume, rattling off strings of purported facts and disguised rhetorical questions, interspersed with promises or appeals to self-interest, and with a dash of the power of suggestion. With the rapid succession of a series of plainly true statement interspersed with false statements, the hearer does not have time to recognise and focus on the false statement. The false statement sneaks in under the radar, so to speak.

It is a technique often used by religious evangelists to whip their congregations into a frenzy; politicians use it in electoral diatribes, or during Question Time in Parliament. One particularly powerful exponent of this technique was Adolf Hitler, as shown in the historical films of the Nazi rallies. But it was not new to Hitler. Military commanders from time immemorial have motivated their soldiers with the same oral diatribes, see for example actor Mel Gibson's (b.1956, age 66) portrayal of the historical figure William Wallace in the movie *Braveheart*.

In some cases, philosophers, poets, writers, and scientists deliberately use obscure or ambiguous language to conceal the fact that they are unable to state their propositions clearly. There is an implicit assumption at play that obscurity equals profundity. Obscurity also discourages close analysis and provides 'wiggle room' for the author and supporters. The alternative is to state the proposition as clearly as possible and then admit the uncertainty in it.

One consequence of the limitations of language is the difficulty it raises for talking about existential and metaphysical issues, which are generally not reducible to a kind of binary analysis (true or false, meaning or no meaning).[44] They are not reducible to the logic of mathematics.

Poetry, literature, art, and theology may be different ways of dealing with existential and metaphysical issues, albeit in a manner that lacks precision. Indeed, some people consider that some areas of philosophy and poetry are almost indistinguishable.

The need for a common language throughout humanity

A discussion between two astronauts on the International Space Station, where one is monolinguistic in Russian and the other is monolinguistic in English, is bound to be a meaningless one. *Star Trek* and *Dr Who*, in different contexts, both found the need to introduce the idea of the 'Universal Translator' in order for different alien species to communicate with each other. China is presently in the process of enforcing Mandarin in preference to any of the other various different Chinese dialects, as the sole official language that all people in China must learn.

There is an obvious advantage to people being able to communicate directly with each other and not relying on a translator, no matter how well-intentioned and competent the translator may be. Each time a statement passes through a translator there is an increased risk of some translational error.

The Olds Testament of the Christian Bible contains an explanation for the lack of a common language, in the story of the Tower of Babel found in Genesis 11:1-9. In this story, men had set about building a large tower that was intended to reach to the sky. God saw this as a challenge to God's authority over men and so deliberate destroyed the tower and scattered the tribes of humankind across the Earth. At the same time God, imposed different languages on the different scattered tribes. This tactic seems to have worked; in that it destroyed any unity in humankind ever since.[45]

Humans now need to reverse the Babel tactic and adopt one of the existing languages as a common language. It probably does not matter which language is chosen as long as it is a living language, that has evolved to meet the needs of changing technology and culture.

But even this simple choice causes conflicts between nationalist tribes, whereby each tribe insists that its language be the common language. Hence the failure of the United Nations to reach consensus on the choice. Viewed from the perspective of the aliens, this is a readily solvable problem and the tribal impediment to the solution appears to be childish and churlish. At the end of the day, the purpose of a language is to communicate, but humans have reduced the choice of the language to one of power politics among parochially motivated tribes.

Communications with the aliens

Human speech, writing, or sign language will of themselves not be of any use in communications with the less developed aliens. They may even constitute a barrier because of the content, the language conditions and the constraints the human cognitive processes, which will almost certainly not be consistent with that of the aliens concenred.

Advanced aliens would have the capacity to communicate with humans by communications implanted directly into the human brain, at either the conscious or subconscious level. This would be interpreted by humans as thoughts, visions or dreams, or perhaps as internal voices. Humans would not know that these thoughts were of external origin. This scenario, featured in the classic movie *Blade Runner*, showed humans as incapable of differentiating between real and implanted memories.

There have been a range of humans in history who have introduced new and innovative ideas that have initiated a paradigm shift in the intellectual history of man.[46] Some of these ideas were incidents of direct alien intervention in the manner described above. Then the obvious question for humans is why do the aliens not reveal their existence and immediately intervene in human intellectual and moral evolution. The answer is simply that the aliens have no intention of intervening and taking control of human evolution, for the same reason as embodied in the so-called 'Prime Directive' of *Star Trek*: to monitor but not disclose or intervene in any overt manner, other than to save. [47]

Humans are not privy to the thoughts of the aliens any more than they are to the thoughts of other humans. If aliens have not revealed themselves, then there must be a reason for it. The fact that humans do not know that reason does not mean that the reason does not exist. In the end, it could be that the purpose of the universe is to create and nurture independently evolving intelligence, that will eventually be elevated to a different mode of existence within the universe. The aliens may only be the agent of that purpose.

In this case, it is entirely wrong to see aliens as a threat to humans. Not only are they not a threat but we are something akin to humanity's metaphorical cosmic 'fairy godmother,' and the producers and directors of *Star Trek* and *Dr Who* may be guilty of defaming us.

The fourth fundamental proposition: aliens are not a threat to humans, but are fellow benign participants in the evolutions of the Universe.

Part 9: Priorities for Humankind's Immediate Future

Protecting the Earth

There will be no intermediate or far future for humankind if the Earth perishes; everyone and everything on the Earth will perish with it. The wealth of an individual will be immaterial, as will be their status in society.

To preserve its present and its future, humankind must, as far as possible, protect the planet and all upon it, from a plethora of risks, including asteroids, solar flares, coronal mass ejections, free-floating black holes, destruction of the environment, destruction of other species, geological and volcanic activity, pandemics, the rise of extremists, wars, and all the social injustices that one human tribe inflicts upon another.

The risks arising from environment destruction and climate change are now at the fore policy, but this was not always so. As long ago as 1798, the poet and philosopher, Samuel Taylor Coleridge (1772–1834), in his lyric poem *The Rime of the Ancient Mariner*, warned of the dangers of man at war with nature. In the poem, an albatross guided the mariner and his ship through the perils of its voyages. But one of the mariner's crew could not help but engage in gratuitous violence and shot the innocent albatross.

As a result, a range of disasters descended upon the mariner and his ship, and no crewmember but himself survived. Before his death, the crewmember that killed the albatross was forced to hang the dead albatross around his neck. The albatross was a metaphor for the magnanimity of nature towards humanity. The dead albatross hung around the neck symbolised the burden that humankind will need to bear, if it goes to war on nature. Naturally, Coleridge's message went unheard.

Humans at war with nature also includes war on the other species on the planet. Humans do not kill only for food. Most life forms on the planet perish as part of the pursuit of profits and through human overpopulation. Some humans kill just for the pleasure of it, or for something they call 'sport.'

Just imagine the fate of humanity if the aliens decided to treat humans and the Earth in a similar manner to the way humans treat each other, the animals, and the Earth.

Humans would feel indignant at the unfairness of the aliens, but the same humans have no conscience about their behaviour towards other vulnerable people, other species, or the Earth itself.

In one episode of an Australian comedy show, two men with high-powered rifles were out at night in a utility truck with spotlights, engaging in the sport of shooting kangaroos. In the show, a kangaroo called 'Skippy' was also armed with a high-powered rifle and was propped up behind the fork of a tree. When the two men fired at Skippy, Skippy fired back. The two men then retreated to their utility and drove to the local police station to report a dangerous and maniacal kangaroo. The men simply assumed the right, as humans, to shoot Skippy and did not allow or expect that Skippy had any right to self-defence. In their minds, the rules of the sport were simple, and Skippy's role was to be killed.

As another example, one of the ever-present dangers to the Earth are the various forms of flares from the Sun, including electromagnetic pulses and coronal mass ejections, both of which are capable of frying electrical circuitry and computer chips. In the late 19th century, the Earth was hit by a large coronal mass ejection, known as the Carrington Event, which took out the telegraph system in North America and elsewhere. The overall damage was minimal because there had been little development in the electrification of cities, machinery driven by electric power, and the use of motor cars.

The Earth was hit again in 1989 by a small ejection that took out the electricity in some eastern parts of North America, and Australia suffered widespread power losses during Easter 2022.

Large solar flares and ejections constantly occur on the Sun. But the Earth is a small target when viewed from the Sun, and so it is not likely that it will be hit again within any particular period of time. But it is a certainty that it will be hit again sooner or later.

If a large flare struck the Earth today, it could take out all the satellites, the overhead power systems and anything running on computer chips or circuitry. There would be massive power surges. All communication systems including the banking system would fail; there would be no electricity for heating, cooling, running engines, driving petrol pumps, or running industry. Cars, trains, boats other than sailing boats, and aeroplanes would not function.

The countries of the Earth would be reduced to conditions akin to a collection of 19th century agrarian communities. But the Earth at that time did not contain eight billion people, and the communities were not megacities. Most people now, including farmers, do not have the relevant agrarian knowledge necessary for survival under these conditions. Cities would be mostly deserted. The Earth could descend into a new Dark Ages that could last for centuries.

Some space agencies are taking the issue seriously and have positioned several solar satellites in orbit around the Sun to monitor and study solar flares. There are some terrestrial stations that monitor the impact of flares on the ionosphere and the interruption to radio communications. But these measures are about detection, not prevention of the consequences of the strike.

Even if a flare is detected by a solar satellite and signalled to the Earth, the signal to the Earth from the Sun takes eight minutes to arrive. An electromagnetic pulse will arrive at the same time and a coronal mass ejection will arrive within 30 minutes after that. There will not be any serious opportunity to shut down the satellites, the electricity grids, all the various engines and take other necessary steps.

Beyond the solar observatories, almost no steps are being taken by the governments and commercial interests to address this threat. Nations continue to launch vulnerable satellites, decline to localise, bury, and harden their electricity grids, or to build terrestrially based communication backups. The weaponization of space continues apace. It is all a product of the politics of profit for which the people of the planet will

In Defense of the Alien

eventually and collectively pay the price. In the event of a failure to adopt change, this is a near certainty.

Expanding into the solar system

In the short term, humans could colonise the solar system the Earth belongs to. The solar system has almost limitless supplies of water and hydrocarbon for fuels[48] that are potentially accessible to human. The asteroid belt is mostly ice water; oceans the Moon and Mars have water, and at least two moons of Saturn and Jupiter have of water greater than those on the Earth. The gas giant planets, Jupiter and Saturn, and the ice giant planets, Uranus and Neptune, each have atmospheres composed of hydrocarbons[49] as does Jupiter's moon Titan. Helium 3,[50] the rare earth metals, and other metals are accessible and abundant.

Colonising the solar system would be a staged process with lessons to be learnt at each step. It will be a huge and hopefully unifying undertaking for the human species. There is currently some cause for hope in the form of the nascent Artemis Program that is now well underway, to colonise the Moon and use it as a springboard into the solar system. But even now, needless geopolitical posturing has put the future of the program at risk.

Reliance on hope can be risky, because hope can act as an anaesthetic to action, and as de facto permission for passivity and procrastination. A contrary view on hope was put forward by the American philosopher, Harold Zinn (1922–2010):

> To be hopeful in bad times is not just foolishly romantic. It is based on the fact that human history is a history not only of cruelty, but also of compassion, sacrifice, courage, kindness.
>
> What we choose to emphasize in this complex history will determine our lives. If we see only the worst, it destroys our capacity to do something. If we remember those times and places—and there are so many—where people have behaved magnificently, this gives us the energy to act, and at least the possibility of sending this spinning top of a world in a different direction.
>
> And if we do act, in however small a way, we don't have to wait for some grand Utopian future. The future is an infinite succession of presents, and to live now as we think human beings should live, in defiance of all that is bad around us, is itself a marvellous victory.
> [Quote by Howard Zinn:]"to be hopeful in bad times is not just foolish... (goodreads.com)]

To date, when people have left the Earth, they report seeing that it is a small and fragile blue orb, the 'Spaceship Earth,' sailing in a dark sea of space on the rim of the blinding Sun.

In Defense of the Alien

For most of these people this experience evokes empathy for the planet and all who sail upon it.

Perhaps if enough people from diverse backgrounds have this experience, it may progressively overcome the parochial 'ant mound' attitude and evoke a similar, but greater, effects than did Carl Sagan's (1934–1996) 'pale blue dot' seen from the spaceship *Voyager*, or the 'Earthrise' photograph taken from the Moon.

Earth from Voyager - Wikipedia]

Earthrise - Wikipedia]

The path for the short-term future

The present endeavours by humans of good will to find political, economic, and ethical solutions to humankind's existing problems take the present state of humanity as a given, with all its primitive evolutionary features discussed throughout this book. They do not posit any significant change on this front. For this reason, these endeavours are just tinkering around the edges of the problem. They ignore the 'elephant in the room,' presumably in the belief that they cannot do anything about it—it is assumed that human nature will remain what it is.

Some politicians and international agencies do struggle with finding a way of encouraging the greedy, the power hungry, the misogynistic, the neurotic—people with all manner of other subconscious drivers and those from the disparate tribes, to come to a reconciliation or agreement on a common approach to humanity's differences and disputes. They attempt to discourage or at least limit wars and other forms of violence. These endeavours mostly fail because of the 'elephant in the room.'

In the near future there appears to be little prospect of any significant change in human nature and so the best that can be expected is a more successful tinkering. Change will be slow—unless, of course, one of the catastrophic events referred to above occurs, and then change will be sudden and probably disastrous.

In the near future humankind should seek to take small steps to attempt to change human nature, mainly through education on a global scale. It could set up alternative role models to the greedy entrepreneur or corrupt government official, discourage false information and manipulation of populations, encourage new metrics for personal success, rein in the excesses of unproductive activities and limit them to their useful domain, encourage the use of a common language and the common pursuit of 'man in space,' and set up a world court and police force with plenary global jurisdiction to actively intervene and stop resort to violence as a mode of dispute resolution.

If there were progress, then 'first contact' with aliens might occur. This would be the most historic happening in human history, in religious terms akin to the spiritual redemption of humankind, the other species of the Earth, and the Earth itself. Each might then enter a new era of evolution towards their respective teleological ends.

One good reason for humanity accepting the challenge of change referred to above is self-interest, because a failure to do so will probably lead to an ensuing cyclical Dark Ages, and then to the eventual extinction of humanity consistent with Bertram Russell's prediction and lament.

[1] *Life's Solution: Inevitable Humans in a Lonely Universe.* Simon Morris, (2009) Cambridge Press.

[2] *Hyperspace*, by Michio Kaku. Anchor Books, 1994. Page 285.

[3] There are thought to have been five major extinction events on the Earth, the latest being only 65 million years ago. But there have been other less severe partial extinctions in the form of ice ages, 'snowball' Earth events, and other climatic disasters.

[4] Alfred Lord Tennyson's *In Memoriam A. H. H.*, 1850.

[5] 'Moore's Law is the law that says that the capacity of computers will consistently double each two years. The analogy is that human evolution accelerates as humans acquire more knowledge and capacity to control or modify their environment.' *Hyperspace*, by Michio Kaku. Anchor Books, 1994. Page 29.

[6] 'Mass' is a reference to matter composed of the known particles of the Standard Model of particle physics. These are the particles that make up the physical, observable universe, but does not include those forms of energy that are massless. Mass is formally defined as a property that interreacts with the Higgs Field.

[7] Often explained in the form of the non-paradoxical 'twin paradox,' and was well illustrated in the 2014 movie *Interstellar*.

[8] Commonly called the 'proper time.'

[9] This is the term commonly used to argue that humanity is not likely to have a personal presence in space and does not have the capacity to communicate with aliens because of the sheer time required for travel and communications.

In Defense of the Alien

[10] Of the Standard Model of particle physics.

[11] Graphic illustrating the Drake Equation:

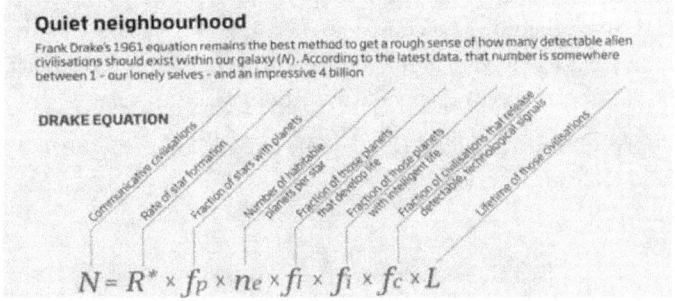

$$N = R^* \times f_p \times n_e \times f_i \times f_i \times f_c \times L$$

[12] If our solar system is young compared to the rest of the universe, and if interstellar travel is achievable given enough time, we should have been visited by aliens already, so where are they?

[13] Science fiction writer Arthur C Clarke stated, 'Any sufficiently advanced technology is indistinguishable from magic.'

[14] The temperatures in the universe range from minus 273 degrees Celsius in empty space, to millions of degrees Celsius near stars. Human can only survive in the small range of 0–50 degrees.

[15] Including breathable atmospheres, edible food, drinkable water, and the capacity to generate usable energy forms.

[16] By 'soul' it is not intended to imply the existence of a separate entity in the brain in the traditional religious sense. It is used as shorthand reference to the collection of the features of human programming.

[17] For example, the contrary view has recently been strongly put by Iris Murdoch (1919–1999) and others (the 'moral realists'),

particularly after witnessing World War II. Other moral philosophers drew different conclusions from their experience of that war. One group of philosophers called the Existentialists concluded that life as such had no meaning, other than what the individual accorded to it. It is also a form of relativism. This choice is said to give rise to existential anxiety in the individual.

[18] Georg Friedrich Hegel (1770-1831) and his followers generally hold the contrary view.

[19] Which are:

1. Wisdom. Wisdom, or prudence, is the capacity to make sensible decisions and judgments based on personal knowledge.
2. Courage. Courage, or fortitude, is the ability to confront fear, intimidation, danger, difficulty, and uncertainty.
3. Moderation. Moderation, or temperance, is the quality of practicing self-restraint and self-control.
4. Justice. The bond that unites the other three virtues is justice. Justice is the quality of being fair.

[20] Christianity derives three theological virtues of faith, hope and love/charity (1 Corinthians 13) as well as the attributes referred to in the Sermon on the Mount.

[21] The greatest good for the greatest number, or the 'calculus of good'.

[22] 1 Corinthians 13.

[23] Aristotle famously wrote 'the unexamined life is not worth living.'

[24] *The Trial* and *The Castle*.

[25] There were early forms of hominid creatures such as Homo Erectus or Neanderthals that go back further, to about 200 million years.

[26] Maslow....

[27] *Fable of the Bees*, Mandeville.

[28] Goethe (1749-1832).

[29] H.G. Wells made this point in his famous novel *The Time Machine* and the relationship between the Morlocks and Eloi.

[30] See https://harrypotter.fandom.com/wiki/Dudley_Dursley%27s_eleventh_birthday

[31] 'The Selfish Gene' Richard Dawkins 1976

[32] https://abcnews.go.com/US/oldest-gay-man/story?id=13320808

[33] https://www.justice.gov/sites/default/files/criminal-hrsp/legacy/2011/06/06/10-30-92mengele-exhibits.pdf.

[34] 'Ode to a Nightingale,' Keats.

[35] WWII, Korean War, Falklands War and Ukraine War.

[36] E.g., WWI, First Gulf War, Suez War.

[37] E.g., Vietnam War, Second Gulf War, Afghanistan, Iran/Iraq war and the American Civil War.

[38] On this premise they sought to take control of the early church (post-legalisation by Emperor Constantine) with all of the privileges that fell to bishops. It led to bloodshed. There were pre-Christian gnostic sects in the early middle-eastern communities, and the Greek and Roman societies.

[39] *The Decline and Fall of the Roman Empire*, Edward Gibbon.

[40] *Strange Gods: A secular history of conversion*, by Susan Jacoby. Pantheon Books, 2016.

[41] https://www.wordfoundations.com/the-bible-and-free-enterprise/ https://www.wordfoundations.com/the-bible-and-free-enterprise/]

[42] The 'unsinkable Molly Brown,' whose life was ultimately made into a Broadway musical.

[43] Academics, social workers, doctors, scientists, artist, journalists, writers, clergy, nurses, carers of various descriptions, and philanthropists.

[44] Analytical Philosophy.

[45] It reads:

1. Now the whole world had one language and a common speech.
2. As people moved eastward, they found a plain in Shinar and settled there.
3. They said to each other, "Come, let's make bricksand bake them thoroughly." They used brick instead of stone, and tar for mortar.
4. Then they said, "Come, let us build ourselves a city, with a tower that reaches to the heavens, so that we may make a name for ourselves; otherwise we will be scattered over the face of the whole earth."
5. But the Lord came down to see the city and the tower the people were building.
6. The Lord said, "If as one people speaking the same language they have begun to do this, then nothing they plan to do will be impossible for them.

In Defense of the Alien

7. Come, let us go down and confuse their language so they will not understand each other."
8. So the Lord scattered them from there over all the earth, and they stopped building the city.
9. That is why it was called Babel because there the Lord confused the language of the whole world. From there the Lord scattered them abroad upon the face of the Earth.

[46] The pre-Socratic philosopher, Democritus, apparently sat on a beach and conceived that the world was made of atoms that combined in different patterns. Anaxagoras (c. 428 BCE) deduced the existence of the stars, planets, and orbital systems. Socrates, Aristotle, Plato, Newton, Galileo, Gutenberg, Bacon, Kant, Einstein, and others may be in this category.

[47] Which is what I interpret the Prime Directive as formulated by *Star Trek* to be.

[48] Hydrocarbons are commonly used fuels on the Earth and in some chemical rocket fuels.

[49] Hydrocarbons are presently common fuels. Petrol and natural gases are forms of hydrocarbons.

[50] Helium 3 is an isotope of helium, that does not occur on the Earth but is plentiful on the Moon and other places. It is likely to form the basis for future nuclear fusion power and propulsion systems.

www.ingramcontent.com/pod-product-compliance
Lightning Source LLC
LaVergne TN
LVHW011710060526
838200LV00051B/2849